Ecological Identity

Also by Timothy Goodwin

Within These Woods

Ecological Identity

Finding Your Place in a Biological World

By Timothy Goodwin

Riverfeet Press
Bemidji, MN

Riverfeet Press
Bemidji, MN
www.riverfeetpress.com

ECOLOGICAL IDENTITY
Finding Your Place in a Biological World
by Timothy Goodwin

Non-fiction
Ecology
ISBN-13: 978-0-9963094-5-5
ISBN-10: 0-9963094-5-4
LCCN: 201946034

Edited by Tracy Goodwin
All illustrations by the author
Cover design by Timothy Goodwin
Author photo by Tracy Goodwin

This book is for the teachers and students at Shattuck-St. Mary's School both past and present, in particular, Terri Dineen, Leah Inman, Greg Simons, and Jennie Sorensen.

Contents

Introduction: Ecological Identity

What is your role in biodiversity?

In a dark cave carved out of sedimentary rock in what is now modern France, a man works by the dim light emanating from a crude oil lamp of animal fat burning in the depression of a stone. Outside, the air is cold. Winds sweep from the north across glaciers covering Northern Europe. Inside the cave, however, the temperature is a constant 50 degrees Fahrenheit. The artist's heavy shawl made from thick reindeer hide, which he has shed, lies on the cave floor. He wears a shirt of sewn-together pieces of tanned horsehide. Leggings made from hides protect his legs from the dampness of the cave.

He grinds a pinch of red ochre pigment and charcoal in his mouth, creating a sticky red paste. At his ready is a palette of yellow ochre and umber. These three main colors are "earth pigments" made from the minerals limonite and hematite. He can make black using charcoal or burnt bones, and white pigment from ground-up calcite. Using his hand as a stencil and his mouth as the spray can he spits the red ochre paste onto the cave wall, forming an image of a horse's head. After he finishes with the body of the horse, he uses a mixture of burnt bone and charcoal to paint the black mane. Next, he draws the outline of the horse's body, tail, and legs with a piece of heavy charcoal. The final component of this piece of art is the artist's signature. He places his hand against the wall with fingers spread wide and sprays red ochre over his hand and the surrounding area. Pulling his hand away reveals a negative-image handprint (Figure I.1).

Cave explorers discovered this 17,000 year-old image in the Lascaux caves, along with nearly 2,000 others, in the 1940s. A great deal about the evolution of human culture can be

learned from collections of such prehistoric art. This, however, is by no means the oldest artwork archeologists have found. The oldest cave paintings attributed to *Homo sapiens* were discovered in Indonesia and are estimated to be around 40,000 years old. However,

Figure I.1

the oldest discovery of what appears to be purposefully created markings dates back perhaps 540,000 years and could have been made by a human ancestor, *Homo erectus.*

The human species, *Homo sapiens,* evolved in East Africa possibly as long as 200,000 years ago. Over the next 120,000 - 170,000 years, we spread throughout the globe, living as hunter-gatherers and probably also as scavengers. Our daily routines, populations, activities, social groups and burgeoning culture were directly connected to the ecosystem in which those early humans evolved and lived. The animals we ate, as well as those we feared, directed our patterns of movement and migration. Early humans were also influenced by changes in the climate and weather patterns because they affected the plants and animals upon which we depended for survival. Intimate understanding of the ecosystem in which we evolved and conducted our daily routines was a necessary skill for survival. This was our ecological identity. How an individual lives and how a species evolves is greatly influenced by the natural world in which they are immersed. Ecological identity is a term that I will first introduce here, but that we will continue to come back to. I use it for describing that relationship to the natural world. It involves how an individual (and a species) describes his or her (or their collective) connection to the natural world. This involves understanding ecological principles, environmental issues, and also then one's emotional or possibly even spiritual connection to the natural world.

During the last 40,000 years, our species developed language, art, and spirituality while living in tribes bound by blood ties. A shift in culture and social groupings began as we learned to harvest and store grains; eventually humans developed methods to cultivate our preferred foods and raise animals for meat and as beasts of burden (animals used for work, like oxen). This dramatic shift from continually migrating to where food grew and animals grazed naturally to an agricultural lifestyle marked a dramatic evolution of our ecological identity and forever changed our connection to the land. No longer just another species, we have now spread to become *the* dominant species on the planet—an elevated position now defining our species' ecological identity.

In this book, through the lens of biological principles, while utilizing visualization of walking through an urban landscape, I will guide you through an exploration of one's role as a part of the earth's living systems. This journey begins with an understanding of where one lives.

The portion of the earth system that can support life, from a few feet down into the topsoil, all the way up to the upper reaches of the atmosphere, is the biosphere. The biosphere is divided into five types of biomes: aquatic, desert, forest, grassland, and tundra (Figure I.2). The human species has adapted to live in all of these biomes with different levels of success. As we have colonized more and more of the earth's biomes, the impact of our footprint on the biosphere has increased. All of the biological concepts necessary to build one's ecological and environmental literacy, and therefore foundational to forging one's ecological identity, happen within the context of one of these biomes.

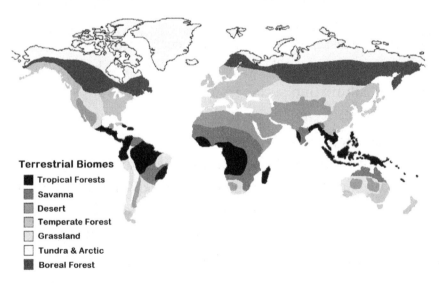

Terrestrial Biomes
- Tropical Forests
- Savanna
- Desert
- Temperate Forest
- Grassland
- Tundra & Arctic
- Boreal Forest

Figure I.2: Terrestrial Biomes

The aquatic biomes include freshwater (ponds, lakes, streams, rivers, and wetlands) and marine (oceans, coral reefs and estuaries). Deserts have little annual rainfall, though not all are stereotypical hot, dry deserts; there are also deserts that are considered semi-arid with cooler nights, deserts along coastlines with cool winters, and even cold deserts that have very short, moist summers followed by long, cold winters with snow, but minimal liquid water available for plants and animals. Forests come in three varieties: tropical forests with year-round, summer-like temperatures and considerable moisture; temperate forests dominated by deciduous trees that lose leaves during winter; and boreal forests dominated by coniferous trees (having needles) and long, cold winters and short, warm summers. Grasslands can be classified as savannas, temperate grassland, and steppes. Savannas are in warm or hot climates and have vegetation that is primarily grasses dotted sporadically with larger trees and shrubs. Savannas tend to have two seasons: a dry and a rainy season. Temperate grasslands are similar to temperate forests in temperature, but often have less rainfall and therefore are not capable of supporting larger vegetation. Steppes have hot summers, very cold winters and vegetation that is usually greater than 12 inches in height. Tundra biomes are extremely cold, with little biodiversity in plants and animals and very short growing seasons. Similar to the tundra are the arctic regions near the poles of the planet and high altitude alpine regions.

You live in one of these biomes, and within that biome are multiple smaller ecosystems, habitats, and microclimates. Understanding your role in the ecosystem, biome, and therefore the entire biosphere is to know your own ecological identity. This requires understanding in two key areas: ecological literacy and environmental literacy. Ecological literacy involves understanding biology concepts and the interaction of living things within their environment. Environmental literacy is more about understanding one's impact on ecosystems. When an individual combines the two concepts, and constructs a scientifically-based understanding of how he or she fits into, affects, and is affected by the ecosystem and the whole earth's biosphere, then he or she begins to form an individual, ecological identity.

One's ecological identity is not static, but instead is dynamic, like ecosystems themselves, and constantly evolving and changing. In fact, our species' ecological identity has evolved as we have evolved. So too can an individual's ecological identity evolve throughout one's life. It is my goal that by the end of this book, you will be able to describe your role in the biosphere and answer the following question based on an understanding of your developing ecological identity: "Then, how ought I to live?"

To help you answer that question, the six chapters of this book address the following thematic questions:

- Chapter 1, Matter and Energy: How do you participate in an ecosystem's flow of energy and matter?

- Chapter 2, Ecosystems: How are you a part of a dynamic ecosystem?

- Chapter 3, Populations: What is your role in a dynamic population?

- Chapter 4, The Balanced Body: How does your body maintain homeostasis within an ecosystem?

- Chapter 5, Genetics: What is the role of genetics in our story?

- Chapter 6, Evolution and Biodiversity: What is the evolutionary history of our biodiverse planet?

I organize and present the concepts here using a systems approach to understanding life on Earth. Instead of breaking living things down into their smallest parts to understand biological life, I prefer an exploration of the interaction between living systems. I will not dismiss the small details, such as the parts of a cell. However, because the focus is on understanding how living systems interact, I will present the small details only within the context of the larger questions.

The interaction of the vast diversity of life on the planet creates incredibly complex systems; yet even an individual organism is made up of complex systems. Living things that are made up of, and are part of, larger living systems have what are called emergent properties. Simply put, the whole is greater than the sum of its parts. This applies to all living systems, be they microscopic, like a single cell, or immense systems like the entire earth system. A biological entity, like the cells in your body, uses energy to organize, regulate, and reproduce itself. In your body, a cell operates independently to maintain itself. However, it interacts with other body cells by exchanging energy, matter, and information. The whole organism, in turn, maintains balance within the ecosystem in which it resides— again, exchanging energy, matter, and information. The ecosystem in which the organism lives is part of larger biome and ulti- mately the entire earth's biosphere.

Video: Homeo- stasis, Autopoie- sis and Emergent Properties *https://vimeo. com/167174459*

All these systems maintain their own balance, and in so doing, contribute to maintaining the ecosystem in which they evolved and reside. Maintaining balance is called *homeostasis*. The hu- man body, or the earth itself maintaining a constant temperature, is an act of homeostasis. Systems consist of many parts working together to maintain homeostasis. Homeostasis is a foundation- al concept to all of the questions posed later in this book.

Western science is just beginning to understand the enormity and complexity of connections between all the earth's systems and organisms. Some scientists prefer a systems or holistic approach to the study of life and ecosystems, requiring an attempt to understand all the connections. Others prefer reductionism, which

Video: Systems Theory Basics https://vimeo. com/167178014

entails understanding a system by understanding all the individual components first. Both approaches have value. Understanding the individual components is necessary, but I agree with scientists who use a systems approach, and will proceed under the following assumptions: that true understanding of a system requires understanding the individual components in relationship to the connections and interactions between all the components of that system; and that truly understanding the details can only happen in the context of understanding the big picture first. Therefore, I present material from a foundation of systems theory and ecological thinking.

The Gaia theory, put forth by James Lovelock in the early 1970s, exemplifies this kind of thinking. He hypothesized that the earth's living and non-living components work together, allowing the earth to maintain its own homeostasis, much like a

living being does. Many people initially misunderstood and dismissed Lovelock, thinking he was proposing that the earth was *actually* a living organism. This was not his hypothesis. Instead, he was proposing that the earth's biological and geological components respond to changes—such as solar energy intensity and tilt of the planet—as feedback mechanisms keeping

Video: Gaia Theory https://vimeo. com/167160918

the earth's temperature relatively constant, a factor necessary for the evolution and survival of living systems. This is analogous to how a living organism maintains homeostasis and internal body temperature, water balance, and chemistry, or how a thermostat keeps a house at a relatively constant temperature. As you read each chapter, contemplate how the processes described in that particular chapter illustrate systems maintaining homeostasis through the exchange of energy, matter, and information. Then consider the role that system plays within the homeostasis of the planetary system that Lovelock named Gaia. And finally, consider what your role is within that system.

As you proceed, it is important to keep the teachings of Western science in perspective. The concept that all components of an ecosystem interconnect is not a new idea; many indigenous peoples understand this and have passed down generations of knowledge about the ecosystems in which their cultures evolved. It was not until recently, however, that Western scientists noticed and began to tap into this source of knowledge and way of understanding. Traditional Ecological Knowledge uses indigenous peoples' collected wisdom about an area or ecosystem. For example, after the 1989 Exxon Valdez oil spill in Prince William Sound, Alaska, scientists used detailed information from the indigenous peoples in the area about the wildlife populations, behaviors, and habitats to establish a baseline of the ecosystem qualities before the oil spill. In another example, Chukotka Inuits provided valuable information about polar bear habitats, which added necessary data used to list the polar bear as endangered under the Endangered Species Act as a result of climate change. This book is not an exploration of Traditional Ecological Knowledge; I bring it up only to highlight that although I present understanding life on Earth in a different sequence than a traditional scientist or teacher may, these really are not new ideas. As you proceed, consider the idea that there may actually be *many* ways of knowing something. This applies to fields beyond biology, ecology, and environmental science as well.

Chapters One through Three are about large systems such as ecosystems, populations, and communities maintaining homeostasis. The focus shifts in Chapter Four to cellular processes, examining the connection between how individual organisms maintain homeostasis through cellular processes and how the individual's homeostasis *fits in with* the homeostasis of the larger systems presented earlier. Chapter Five looks specifically at genetics and how individual organisms store and pass information from one generation to the next within the context of their ecosystem. Chapter Six concludes with an exploration of the diversity of life on planet Earth and how a species' evolving genetic code contributes to that diversity. As an aid to your understanding, there is a companion curriculum with many online resources, such as short video lessons, available at www.exploringbiodiversity.com.

QR codes for supplemental video lessons are embedded in the text when that topic is covered. At the end of each chapter is a QR code link to the web resources connected to that particular chapter and a QR code link to the web resource home page.

Web Resource Home Page
www.exploringbiodiversity.com

Introduction Web Resources
http://www.exploringbiodiversity.com/#!blank/c1cqy

Chapter 1: Matter and Energy

How do you participate in an ecosystem's flow of energy and matter?

1 – Biological Communities

If I were to ask you to imagine walking outside on a sunny day in the "natural world," what is the image that comes to your mind? I would bet that you pictured yourself surrounded by nature, maybe a forest, or a meadow, or some other such bucolic setting. It is, excuse the pun, natural to do this. This need for connection to the natural world may be stored in your genetic code and be the expression of millions of years of "genetic history." Biologist E. O. Wilson put forth a hypothesis, called biophilia, stating that we humans have an innate love for the natural world, and especially animals. When *I* try this exercise, my mind conjures up an image of a northern Minnesota or Wisconsin lake (Figure 1.1). I picture calm waters and a clear, blue sky with the soundtrack of singing birds and rustling

Figure 1.1: Wisconsin Lake

11

leaves in the summer wind. It is a serene setting for me. What natural place is a part of your ecological identity? What are your default images when asked to think of the natural world?

This is too easy of an exercise though. Let's make it more challenging. Instead of in your "natural place," I want you to imagine yourself in the most metropolitan setting you have ever experienced. Picture yourself walking down a busy sidewalk in a large city like New York, Chicago, London, or Hong Kong. Even in the most urban of settings you are still surrounded by life, by biology and ecosystems.

As you take this imaginary walk down this city sidewalk notice the large trees growing through specially made holes in the concrete that allow water to get to the roots. The tree has leaves with three points and seeds that helicopter their way down to the ground below. This particular tree is a sugar maple, *Acer saccharum* (Figure 1.2).

The name, *Acer saccharum,* is this species' Latin name. In the 1700s a Swedish botanist, Carl Linnaeus, invented a system for naming all the plants and animals, and at that time, living things were put into just those two categories. The system he invented uses binomial nomencla-

Figure 1.2: Sugar Maple

ture, or "two-name, naming system." He chose Latin because it was a universal language at that time. Biologists still use this system to name individual species, though now they might use names that just *sound* Latin, but actually are not real Latin words. Often a newly discovered species is named after a Latinized version of the scientist's name or the location where the first specimen was discovered. For example, an extinct species of humans, called *Homo heidelbergensis,* was discovered near Heidelberg, Germany, in 1907. It went extinct about 250,000 years ago.

In the case of the sugar maple, think of the first name, *Acer,* like your family name, so in my case, Goodwin. The second name, *saccharum,* is the name specific to that particular kind of tree (or species), and is analogous to your first name—Tim in my case. There is only one you, and as much as you are unique, every species on the planet is also unique. In this case, the sugar maple is the most abundant of maple trees around the Great Lakes of North America, where I have lived most of my life and what I visualize when imagining "nature." For hundreds of years, humans have used this tree for lumber and for sugar production, which is why it is named for the Latin word for sugar, *saccharum.* Sugar maples live a long time and will not even produce seeds until they are at least thirty years old. Living in this tree that is providing you shade as you walk along the city street might be many different kinds of insects—maybe a tent caterpillar or a maple petiole borer. In addition to the insects are predators that eat those insects, such as spiders, praying mantis, or birds like the American robin, *Turdus*

Figure 1.3: White-Throated Sparrow

migratorius, Northern cardinal, *Cardinalis cardinalis,* or white-throated sparrow, *Zonotrichia albicollis* (Figure 1.3).

Even a busy, concreted-over metropolis like the one I am asking you to imagine is a part of the biosphere. Life is everywhere, using energy from the ecosystem to continually recycle matter.

Each and every day you interact with countless species as part of a greater system. Most people do not understand and recognize the myriad of components interacting to make the system function. Even to many scientists it can at times seem like an almost incomprehensibly complex system. It is a system with one purpose—to maintain itself and its own homeostasis. Within this system, not only does life surround you, but you are connected to all of those living things around you. Whether you recognize it or not, can readily see it or not, we are all part of a biological community. It is here, picturing yourself as a part of that community of life, that we begin our exploration of your role in biodiversity and in the ecosystem in which you live and ultimately within the entire biosphere.

So what then is an ecosystem? An ecosystem is the interaction of all living and non-living things in an area. Living things (or once living, like a rotting log) are referred to as biotic factors. Non-living things are called abiotic factors, like heat and energy coming from the sun, the water in a lake, or the rocks on a shoreline. It is worth emphasizing that an ecosystem is the interaction of *everything,* all of the abiotic and biotic factors within a specific area.

Let's begin with the living portion of the ecosystem: the community. In popular language, the term "community" describes a group of people living together. A community might be defined simply by geography, such as all the people living in a town. Sometimes a community is intentional, such as all the people attending the same school or on the same team. The latter definition usually implies a common, or shared, purpose.

The term community, in a biological sense, simply means all the living things in an area, or within an ecosystem. Though technically accurate, I find this scientific definition a bit limiting. I prefer to think of a biological community as more like the common definition—a group with a *shared purpose* more than the simple idea of a group of organisms living together. The shared purpose of the community within an ecosystem is to maintain the ecosystem's homeostasis. However, we want to be careful not anthropomorphize this idea, meaning to give the system and the organisms human qualities. Unlike the shared purpose of a group of people who chose to be on a team and work together toward a common goal, the organisms did not choose to be in this community, nor are they *consciously* working toward a common goal. That said, they all did evolve within that ecosystem and do contribute to the homeostasis of that ecosystem.

Consider a ballet, a theatrical play, or a carefully executed play by a sports team. All members of those communities work together, relying on each other to achieve the greater objective, yet all execute their roles independently. Similarly, all living things participate in a complex dance, as choreographed as any sophisticated ballet. All the living things fulfill their roles and work in connection with one another, even if they do not know it, to maintain the homeostasis of an ecosystem, and therefore contribute to maintaining the homeostasis of all of planet Earth's biosphere.

Video: Biological Communities
https://vimeo.com/167464455

2 - Energy In

Let's get back to our imaginary metropolitan walk, with you continuing down that busy city street. You pass in and out of shadows cast by the buildings. The morning air has a bit of a chill to it and there is a stark difference in temperature when you move from the shadows to the sunlight. Imagine the sun on your face. It warms you and feel goods, especially on this cool morning. But beyond warming you and providing some vitamin D, it provides no direct benefit to your body. You, of course, cannot absorb the energy directly from the sun. You are a consumer, meaning you get energy by eating other organisms. Most living things are consumers, except those that absorb energy directly from the sun like plants and some photosynthetic bacteria called cyanobacteria. The first topic in our journey to understanding how energy flows through the ecosystem is how energy gets from the sun, into the ecosystem, to consumers such as you, and then eventually to the decomposers.

The sun has so much mass that at its center a type of nuclear reaction called fusion occurs when hydrogen and helium atoms are forced together, or *fused*. When atoms are fused, especially atoms that are of different size, such as helium and hydrogen, the reaction produces great amounts of energy. This energy leaves the sun in various forms of light, from ultraviolet light, to visible light, to infrared light. This range of light is called the spectrum (Figure 1.4), and even though most of it is not visible to our human eyes, the energy is striking you and all of the rest of the

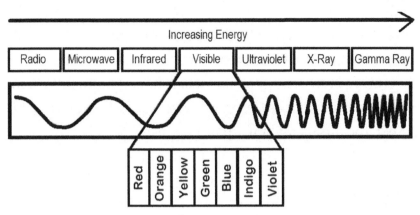

Figure 1.4: Electromagnetic Spectrum

earth nonetheless. To understand your role in the transfer of energy through the earth system, it is necessary to understand how energy is collected by the biological community, what is done with that energy, and where it goes when living things are done using it.

Walking down our imaginary city street requires your body to use energy. Before you set out on this walk imagine the breakfast you ate—maybe an egg and toast, or bowl of cereal. You and all organisms, except plants, are heterotrophs, meaning you get your energy by consuming other organisms. Your body is powered in large part by the complex carbohydrates in the food you eat. Complex carbohydrates are large molecules made up of many sugar molecules strung together. The energy within the chemical bonds of molecules, such as complex carbohydrates, eventually traces back to origins in the sun. So how is energy stored in the bonds holding atoms together to form these molecules?

A bond between two atoms is made by the interactions of the electrons of those atoms. Atoms can either share electrons with one another, or electrons can be transferred from one atom to another. In either case, energy is involved because there is a change happening. Therefore, the act of forming or breaking a bond between atoms requires energy. Because of this, we say living things store energy as molecular bonds. It might be more accurate to think of it as the energy is stored in the *relationships* between the atoms making up molecules.

But what does it mean to store energy in these bonds? To better understand this concept requires a little detour from our biology discussion and into the world of physics and thermodynamics. It is easy to think of energy as something that is produced, used, and then lost since our most common interaction with "energy" is in heat or power from burning something. We can feel the heat dissipate over time, or witness that when we stop burning fuel in our car, the car stops moving. According to the first law of thermodynamics, however, energy can be transformed from one form to another but it can never be created or destroyed. The implication of this law is that there is a fixed amount of energy in the universe—always has been, and always will be. The energy can be changed from light energy to heat, to chemical energy, and so on. The amount of energy in smaller systems can fluctuate, say in a cell, a body, or an ecosystem, as energy might flow from one system to another. However, through all those changes, the total amount of energy in the universe remains constant. The reason I mention this first law of thermodynamics now is because it applies to the energy originating in the sun, captured by the earth, and used by all of the living things as they live and grow.

The second law of thermodynamics is slightly more complex and abstract. It is important for our purposes here to understand how energy flows through a system to support the growth, evolution, and homeostasis of the systems we are going to explore—the earth system, living systems like a plant, your body, and even individual cells. Within the earth system, the plants uses the light energy to make complex carbohydrates. Unlike the heterotrophs, plants make their own food and are therefore referred to as autotrophs. Plants do this by transforming light energy into chemical energy. Your body can then transform that chemical energy stored in the bonds of the carbohydrates into movement, such as flexing leg muscles or autonomic functions such as the beating of your heart. With each transformation, the energy becomes less complex and usable to another system until it is finally in its simplest form: heat. This is the essence of the second law of thermodynamics. The amount of energy is constant, but each time it gets transformed, some of the order may be lost, which increases the entropy of the system. Entropy is the measure of a system's order, or more precisely, lack of order. The more chaotic, the higher the entropy. Eventually, the energy used in a system ends up in its simplest, least ordered state, which is heat energy. This concept is foundational to understanding the flow of energy through a living system, like an ecosystem.

Video: First and Second Law of Thermodynamics
https://vimeo. com/167745748

Upon first look, it might appear that the evolution of life, which usually results in things becoming more ordered and complex, seems to break the second law of thermodynamics that says with each energy transfer systems should get more chaotic and entropy should increase. In fact this is often an argument used to try and "disprove" evolutionary theory. This argument is a specious one, however. Within the homeostasis and evolution of living systems, the energy is used by the system to increase its sense of order, and therefore decrease its entropy, while decreasing the order of the energy and thus increasing the entropy of the energy fueling the homeostasis and evolution of that living system. So while the smaller system may be *increasing* in complexity—say, mammals evolving eyes—the total energy in the overall system remains unchanged and the entropy of the entire system increases as the smaller systems transform stored energy into heat energy, thus still abiding by the second law of thermodynamics. The process of how this evolution happens is covered in much more detail in Chapter Six.

Okay, now let's connect all of this physics back to energy in the ecosystem. Pause your reading and feel your breath on your hand. It is warm. While you are reading this text, your brain cells need energy, and so they are breaking the bonds between atoms in sugar molecules to do the work necessary to think about and comprehend what you are reading. Thinking is hard work and takes a great deal of energy. In fact, the brain is a significant consumer of energy and requires large-brained mammals such as our species to consume considerable calories each day just to maintain itself. If you were indeed walking down that city street instead of just imagining it, you would have to burn even more sugars to power your legs and all your body's life functions as well as power the brain to navigate where you are going.

The energy you have used while reading this began as nuclear energy in the sun, which emitted that energy as light energy toward Earth. Plants used some of that energy to grow, and then stored the rest as sugars bonded together into complex carbohydrates, most commonly as molecules called starch and cellulose. You got the energy from the plants directly, either by eating food made from plants or by eating an animal that ate the plants. This energy powers all your cellular activities while the leftover energy is lost in the form of heat or stored in your body for later use. That is why you can feel heat when you hold your hand over your mouth and exhale. You are a part of the transfer of energy, and are affected directly by the first and second laws of thermodynamics as your body continuously forms and breaks bonds between atoms to utilize the energy stored in those molecules.

Now remember, energy is not stored *inside* the individual atoms (like carbon or oxygen) you ingest, but in the relationships or *bonds* between those atoms in different molecules like glucose (Figure 1.5). This is one of the most basic examples of the interactions and connections between components of an ecosystem. The homeostasis of living systems emerges out of those interactions and connections. Understanding these connections is elemental to understanding one's ecological identity. The individual atoms of carbon and oxygen and water have properties endemic to their structure, but when these individual atoms are connected through the exchange and/or sharing of electrons new properties emerge because of how they are arranged in relation to one another. Connecting these same atoms in different combinations creates unique molecules with

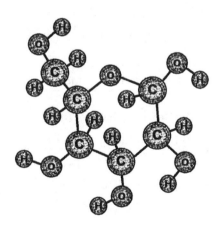

Figure 1.5: Glucose Molecule

unique properties. The structure of a glucose molecule differs from that of a carbon dioxide or water molecule (Figure 1.6), even though they are made from the same component parts. Because they have different structures and different combinations of bonds, they have different amounts of energy and different functions in the ecosystem.

Glucose is a sugar made by plants. It is a molecule of particular importance when discussing energy in an ecosystem. A plant uses the external source of energy from the sun to drive the chemical reaction that makes glucose, which, is the fuel for all life on our planet. As atoms are rearranged, say from glucose to carbon dioxide and water, the energy needed to break the bonds in the glucose releases more energy than it takes to put those atoms back together into carbon dioxide and water. This process produces a net gain of energy. The cell captures this energy to power its activity. Eventually, this energy leaves your body in the form of heat energy. At this point you have utilized as much of this energy as your body can, but this energy that originated from nuclear fusion in the sun is still in existence. It does not disappear from the system.

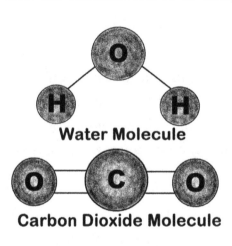

Water Molecule

Carbon Dioxide Molecule

Figure 1.6: H_2O and CO_2 Molecules

Every energy transfer abides by the first and second laws of thermodynamics. However, each of these transfers of energy is actually quite inefficient. With each transformation, much of the energy is lost to the living system as heat—once again, the least ordered and therefore least useful form of energy. Once in this form, it is unlikely to be used by another living organism. It still exists, but in its lowest level of complexity. Therefore, this increase in entropy is an increase in disorder. Energy flows through an ecosystem, allowing the ecosystem to maintain homeostasis, but the energy, though not lost, is transformed and the system as a whole increases in entropy. The details of how energy is transferred between organisms within an ecosystem will be explored in Chapter Two. But first, let's dig deeper into the first stop for light energy entering our biosphere.

3 – Capturing the Energy

The light energy warming you while you walk down our imaginary city sidewalk takes about eight minutes to travel from the sun to you and your cells. How light travels, and in what form, is a tricky concept; for the purposes of our investigation, think of light as traveling in little packets of energy called photons.

All living things are made of cells. And it is through the functioning of cells that an organism's life functions occur, including the capture and use of energy. Therefore, before we proceed, some overview of the functioning of cells is necessary. At this point, we will just look at two kinds of cells, "typical" plant and animal cells (Figure 1.7). No plant or animal has a cell that is "typical," as each is specially modified to perform a certain task, but it is helpful to begin with understanding a cell as if it could

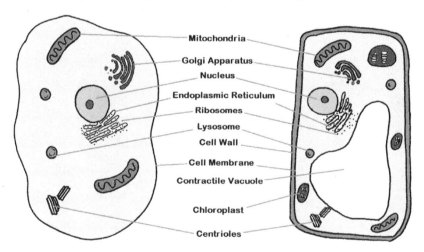

Figure 1.7: Typical Animal and Plant Cell

equally do all the things a typical cell does. Know that in reality, no cell is a "jack-of-all-trades" and can do everything equally. Each is specialized to do a specific task in its contribution to maintaining the body's homeostasis, such as store information, carry oxygen, burn energy, secrete digestive enzymes, and so on.

The main difference between plant and animal cells is the presence of chloroplasts and cell walls. Plants have both of these while animals do not. The chloroplast absorbs the sunlight to make food for the plant—more about that in a bit. The cell wall provides the plant with structure, sort of like the skeleton of an animal. Beyond that, the "typical" cell for plants and animals is similar.

On your imaginary walk you notice a car dealer. I like to think of the cell as like an auto plant assembly line. The main office of the auto plant is like the nucleus of the cell. Within the office are the engineering drawings for the car being made by that auto plant. While the auto plant stores the information as engineering drawings, the cell uses of the genetic code on the DNA. The nucleus of the cell directs the activity of the cell. Cell activity happens through the use of proteins. The auto plant produces a car as the final product while the cell produces proteins needed by the cell and/or the larger body of which it is a part. The original engineering drawings for the car being produced is a closely guarded industrial secret and therefore is not allowed out of the main office. Like those drawings, the DNA code is unique and valuable. Therefore, the nucleus makes a copy of the DNA code and sends a copy, called messenger RNA (mRNA) out of the nucleus. The copy of the code is transferred through the endoplasmic reticulum, which is like the mail system or hallways of the auto plant, to the assembly line. The assembly line is the ribosome, which reads the genetic code and assembles a protein. That protein is the product. It might be an enzyme to help digest food, or maybe a replacement part for a cell component. Once assembled, the protein is then packaged by the Golgi apparatus, which is the packaging and shipping portion of the cell. From here, the cell transports the newly assembled protein to another location in the cell, or out of the cell entirely. This is where the cell membrane enters the picture. The cell membrane is what the cell uses to control what moves in and out of itself.

In addition to producing a product such as proteins, new cell parts, or chemicals needed throughout the body such as hormones, the cell also has to maintain itself. All living things need food, water and energy. The vacuole is the storage room of the cell. Plants primarily use a contractile vacuole to store water, while some animal cells may have this as well as a food vacuole. The mitochondrion is the organelle inside of the cell that is the generator for the factory, providing all of the energy to maintain the cell by burning the food the cell eats, or makes in the case of plants. Life inside a cell, like life inside a busy factory, is messy. The lysosome is the waste disposal and recycling center of the cell. It breaks down waste, recycles cell parts no longer functioning, and cleans up inside the cell, like the janitorial service for a factory.

This is a very brief overview of the cell and its component parts called organelles (tiny organs). As you proceed through the rest of the book, the functioning of the organelles will be explored in more detail, when needed to understand how the functions specific to those organelles help the cell, organism, or ecosystem maintain homeostasis.

The cell functions applicable to this chapter's topics revolve around the cell capturing and using energy. The first stop for the energy on its journey through our biosphere is in the cells of a plant's leaf. Photons from the sun strike the leaf of a plant. This could be something as large as a tree, or as small as a woodland wildflower such as a large-leafed trillium, *Trillium grandiflorum* (Figure 1.8). The plant uses a complex molecule, called chlorophyll, stored in a structure inside its cell called a chloroplast. Plants capture light energy here and begin the process of converting it into chemical energy in the form of sugars, mostly in the form of glucose. How the cell captures light energy will be described shortly, but before we can explore that process, we first need to understand the overall function and role of chlorophyll in the plant.

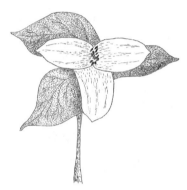

Figure 1.8: Large-Leafed Trillium

The plant is green because the chlorophyll molecule is a green pigment. There are other pigments as well, but they are less abundant than the green chlorophyll. The purpose of this pigment has nothing to do with camouflage, like certain colors often are for animals, but instead is a by-product of the plant's adaptation to capture the sun's energy and produce glucose. As chlorophyll absorbs the sun's energy, the sunlight constantly breaks it down. Therefore, the plant has to continuously make new chlorophyll molecules, which requires a lot of energy on the part of the plant. In the fall, when the days shorten and there is less sunlight, it is no longer worth the investment of energy to absorb the sunlight at the cost of the chlorophyll. Therefore, plants in temperate regions stop making new chlorophyll. While the chlorophyll gets used up in the leaf, the red anthocyanins and yellow-orange carotene pigments remain. The leaves then "change" to their fall colors of oranges, yellows and reds. These other colors were always in the leaf, but were covered up by the abundance of green pigment. With the green pigment decayed and depleted, it is time for the other colors to shine, literally, as the red and orange parts of the spectrum in the sunlight are now reflected from the leaf back to your eyes more than the green.

Most of the light energy coming from the sun is in the form of gamma, X-ray, ultra-violet, infrared, and radio waves. The visible light allowing us to see the leaves of a tree is only a small portion of the energy from the sun. This is the familiar rainbow of red, orange, yellow, green, blue, indigo, and violet (ROYGBIV). There are other colors present, but these are the only ones the human eye can see. Some animals can see fewer colors than we can, while some can see many more. Within the visible part of the spectrum are the colors plants absorb and convert into chemical energy.

Green plants appear green to human eyes because the majority of the plant pigment is green chlorophyll. Green (and some of the yellow) from the spectrum reflects *off* the leaves and strikes the retina in the back of our eyes, making the leaves appear green. The plant absorbs the rest—the ROY and BIV portions of the spectrum. Of all the sun's energy striking the earth, the plants only capture about one percent. The rest reflects back into space or is absorbed by abiotic surfaces like concrete, oceans, and rocks.

The fact that the chlorophyll molecules reflect the green part of the spectrum is simply a by-product of the molecule's shape. Because green pigment in plants is the adaptation that allows plants to absorb sunlight and convert it into chemical energy, green is the background color for much of the ecosystems of the world, and therefore, a great many animals evolved to fit into that green backdrop.

Remember, the purpose of all this chemistry and physics is to produce sugars, not to make the world a nice shade of green or provide hiding places for insects. As I have explained, one of those sugars is glucose ($C_6H_{12}O_6$), a simple sugar made up of 6 carbon atoms, 12 hydrogen atoms, and 6 oxygen atoms. It can be either a string of six carbons with hydrogen and oxygen molecules coming off the sides, or a ring with six sides, each corner occupied by a carbon atom except one with an oxygen atom. There are other forms of sugars produced, but for our discussion of energy transfer through the ecosystem, we will focus on this one simple sugar. What do you notice about the atoms that make up glucose? Pretty common stuff isn't it? Where do you think the plants get these atoms and molecules? Where in the ecosystem were those atoms before the plant assembled them into the glucose molecule?

To answer these questions, let's look more closely at the chemical reactions that happen during photosynthesis when a plant makes sugars. Chemical equations are used to represent the changes that occur when atoms are rearranged. Often, energy is also shown in the equation. The overall chemical equation for photosynthesis is:

$$6CO_2 + 6H_2O + \text{Light Energy} \rightarrow C_6H_{12}O_6 + 6O_2$$

The left side of the equation represents the starting molecules and the right side of the equation represents the concluding molecules. The arrow represents the conversion of matter from one form to another.

Photosynthesis occurs in the chloroplast of the plant cell and involves many complex chemical steps (Figure 1.9). Understanding energy in the ecosystem does not require memorization of these complex steps, but instead understanding two processes: the light reactions and the Calvin- Benson cycle.

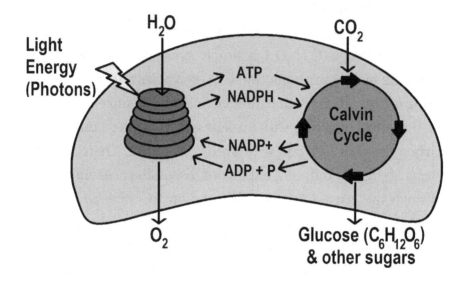

Figure 1.9: Photosynthesis Overview

The light reactions capture the energy from the sun through the movement of electrons removed from other atoms and single protons without an electron called hydrogen ions, written as H^+. This energy is not in its final form, however. The Calvin-Benson cycle describes how the energy is used step-by-step by the plant to assemble the final product of photosynthesis: a glucose molecule.

Before we look more closely at this process, a quick review of atomic structure is needed. Consider a hydrogen atom (Figure 1.10). An atom is represented with an electron orbiting around the nucleus of that atom, which in the case of H^+ consists of

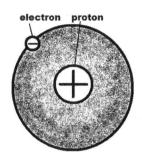

Figure 1.10: Hydrogen Atom

a single proton. This model is not entirely accurate, but in the end, it is the best model scientists have created that does not require complex math to understand.

Now let's take a closer look at the light reactions. During these reactions, photons (remember: little "packets" of light energy) strike the chlorophyll molecule that is in the chloroplast inside the leaf cell. Chlorophyll captures light energy when a specific light wavelength matches the energy required to literally move an electron on the chlorophyll molecule further away from the nucleus of the atom and to a higher energy state. Imagine the photon as a little ball from the sun. It hits the electron on one of the atoms within the chlorophyll molecule and knocks that electron further away from the nucleus. Moving the electron in this way requires an input of energy from the sun. Since energy can neither be created nor destroyed, the energy it took to move the electron is effectively stored in the *movement* of the electron. When an electron is moved away from the nucleus it is said to be "excited." This electron is then transferred through a chain of reactions called the electron transport chain. This transfer of the excited electron is how the plant captures the sun's energy. The plant then replaces the electrons lost from the chlorophyll by taking them from hydrogen atoms. This leaves behind hydrogen atoms without an electron, which are now the simple, single, positively charged protons known as hydrogen ions (H^+).

These single, positively charged protons (H^+) are moved disproportionately across a membrane in the chloroplast so there are more hydrogen ions trapped on one side of the membrane.

The more uneven the spread of the protons, the more potential energy is stored. It is like a charged battery. As the protons move back across the membrane, the movement powers a pump that makes energy-rich molecules called ATP and NADPH, which are important, high-energy molecules cells use for many functions, including the finishing of the production of glucose (Figure 1.11). We'll

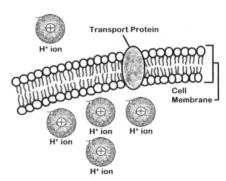

Figure 1.11: Movement of Hydrogen Ions

come back to the structure of these molecules later. For now, let's continue our exploration of the conversion of light energy into stored chemical energy.

Synthesizing glucose to store energy in the bonds is not done in one step. In actuality, the end product of the light reaction is incomplete sugar molecules, each made of only three carbon atoms (along with oxygen and hydrogen atoms). The energy stored in the bonds between the atoms in the energy-rich molecules of ATP and NADPH is now used to power the second process of photosynthesis called the Calvin-Benson cycle.

The Calvin-Benson cycle is named after the scientists who discovered it, Melvin Calvin and Adam Benson. They were awarded the Nobel Prize for chemistry in 1961 for their discovery, and Calvin was labeled "Mr. Photosynthesis" by *Time* magazine that year.

The Calvin-Benson cycle, like the light reaction, is a complex process. It is the cell's way of using the energy from the sun to complete a repeating cycle of chemical reactions that combine the 3-carbon sugars into more complex sugar molecules for storage. It uses the ATP and NADPH from the light reaction, molecules that have unstable but high-energy bonds holding them together. The energy in those bonds is used to rearrange the carbon and oxygen atoms into 3-carbon chains. These 3-carbon chains are then run through a repeating cycle that results in sugars.

Remember, the sugar is the ultimate purpose for this complex process of photosynthesis, which explains how the sun's energy is converted into chemical energy and stored as molecules. But where does the plant get the atoms to make the sugar molecules? Have you ever wondered why plants need to be watered? We all know they do, but it's not just so they simply do not dry out. Look back at the chemical equation for photosynthesis and notice that the source of hydrogen atoms is water. This

Video: Photo-
synthesis
*https://vimeo.
com/167557336*

is why watering plants is important. The plant breaks apart the water molecules to get the hydrogen atoms it needs to replace those used in the light reaction portion of photosynthesis. These atoms eventually end up as part of the glucose molecule. The single oxygen atoms left over quickly bond with one another to form O_2 and then are released into the atmosphere as the oxygen we breathe.

As you visualize your walk down the street, you take a deep breath of the morning air, feeling the coolness fill your lungs. You are inhaling oxygen (along with other molecules) and exhaling carbon dioxide, CO_2 (along with some other molecules). The plants "exhale" O_2. You need the oxygen and the plants need the carbon dioxide. The CO_2 you exhale provides the carbon and oxygen atoms that eventually end up in the glucose molecule.

These molecules have mass and weigh something even though you cannot see them. To imagine this, think back to when you were young and playing in a restaurant play area ball pit. Think of being surrounded by all those differently colored plastic balls. Now, in your mind, shrink those down until they are too-small-to-see plastic balls, each representing a molecule of CO_2, O_2, N_2, or H_2O. Imagine tiny molecules of carbon dioxide floating out of your lungs, then into the air, and into the leaves of a nearby tree. Oxygen, escaping from the leaves, replaces the space vacated in your lungs by the carbon dioxide. The energy stored in the bonds of glucose molecules, made from these common molecules, exchanged between all living things, creates an actual, physical connection between you and the tree you might walk by on our imaginary city street. This real and quantifiable physical connection is never considered by many, but cannot be dismissed as it is an integral component of one's ecological identity.

Glucose is the most basic of the sugars made by the plant, but it can also manufacture fructose. These simple sugars are called monosaccharides (mono meaning one). Two monosaccharides put together, such as one fructose paired with a glucose, makes sucrose (a disaccharide), more commonly known as table sugar. Plants assemble many sugar molecules to form complex carbohydrates called starches (polysaccharides) in order to store energy, much like animals store energy in the form of fat. A potato, carrot or rutabaga is an example of the structure a plant uses to store large quantities of carbohydrates in one convenient location for later use.

Let's boil this down to a simple summary statement. The

result of photosynthesis is this: the inputs of light energy, carbon dioxide and water result in an output of stored chemical energy (glucose) and oxygen (Figure 1.12). We now have energy successfully harnessed from the sun and stored as chemical energy ready to fuel all life in the ecosystem.

Now what happens to it? How does the plant use that energy, and how do the animals get the energy from the plant? Remember that the first law of thermodynamics states that energy can neither be created nor destroyed, but only transformed. The second law states that as it is transformed, some is lost as heat and essentially is no longer usable to us— though still in existence. Where then does that heat come from and where does it go?

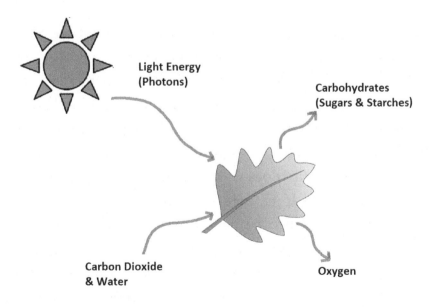

Figure 1.12: Photosynthesis Overview

4 – Using the Energy

I have already introduced the atoms used to make the molecules that store the energy for living things: carbon, oxygen and hydrogen. Recall that during photosynthesis, plants breathe in CO_2 and breathe out O_2. You already know that while sitting here reading this, or walking down a sunny sidewalk, you are breathing in O_2 and breathing out CO_2. The act of breathing is more properly called respiration. Hang on to that term. It is also, I am sure, obvious to you from experience that the faster you are moving, the faster you breathe. How do you suppose this also relates to the rate at which you are using, or burning energy, and what does it mean to "burn" it?

When the plant breathes out oxygen, it does not let all of its oxygen go. The plant also uses it for the same purpose you do—to burn its food. Plants have to eat too, but they eat the food they make during photosynthesis. I use the term "burn" purposefully. For something to burn, what is required? Why does water, a blanket, or a fire extinguisher put out a fire? Anything that extinguishes a fire does so by separating the oxygen from the fuel being burned. Adding oxygen to something is said to "oxidize" it. This means the paired oxygen atoms (O_2) break their bonds with each other in favor of bonding with another atom like hydrogen or carbon. Oxygen is normally written as O_2 because its natural state is two atoms bonded together. When two oxygen atoms share electrons, forming O_2, they share them equally. However, when oxygen atoms share electrons with atoms like carbon or hydrogen, the electrons move *closer* to the oxygen nucleus, which causes a *release* of energy. This process is the opposite of when electrons are moved away from the nucleus of an atom by the input of energy and they become excited. The action of bonds breaking and reforming in different arrangements releases large amounts of energy. This release of energy can be seen in the form of light or flame when a fuel source burns.

Imagine a flammable material such as gasoline, wood, or coal. Whereas glucose has a few carbon, hydrogen, and oxygen atoms and bonds, these fuel sources have hundreds, thousands, even millions of atoms all bonded together. By adding more oxygen to it, the bonded carbon atoms break apart and bond instead with the available oxygen. This releases energy in the form of heat and light—a flame. The waste product is the CO_2 that is released along with the particulate exhaust or smoke.

Now, let's look at how this compares to what is happening in your cells. The same process is happening in all living things. Instead of burning fossil fuels like those in a car's engine, or a piece of wood in a campfire, the cells of living things burn simple or complex sugars (carbohydrates). The result of burning these sugars is waste (CO_2) and heat. Heat energy not captured by the cell to perform work is released into the atmosphere. All of these burned molecules have something in common—they are collections of lots and lots of carbon atoms bonded with oxygen and hydrogen, which were put into that form by plants fueled by the sun's energy during photosynthesis. We call these molecules and the living things made of them, organic. In cells, we call this process of burning fuel *cellular* respiration. Can you see the connection between "breathing respiration" and "cellular respiration"? Because this cellular process of burning the glucose uses oxygen, it is called *aerobic* respiration; "aerobic" means "with oxygen."

Cellular respiration, like photosynthesis, is a complex chemical reaction, most of which occurs in the mitochondria inside the cells. Therefore the mitochondria are sometimes referred to as the "powerhouses" or "energy plants" of the cell. The chemical equation for cellular respiration is:

$$C_6H_{12}O_6 + 6O_2 \rightarrow 6CO_2 + 6H_2O + \text{(up to 36 ATP)}$$

Now, take a second look at the chemical equation for photosynthesis:

$$6CO_2 + 6H_2O + \text{Light Energy} \rightarrow C_6H_{12}O_6 + 6O_2$$

Notice the similarity? Instead of using energy to make glucose, the cell disassembles glucose, resulting in stored energy in ATP molecules and waste products (CO_2 and H_2O). Just as it takes many steps, through the transfer of excited electrons, to build glucose, cellular respiration is a process of many steps, also requiring the transfer of electrons, to dismantle that glucose. Think of it this way: whatever goes up, must come down.

We will focus on and summarize two of the many steps involved in aerobic cellular respiration. The first step is glycolysis. The suffix "-lysis" means to break apart, so we are talking about the disassembly of glucose with this step. Four ATP molecules are produced during glycolysis, but the cell ultimately gains only two ATP from this process because it requires a bit of energy— two ATP—to kick-start the process. This happens outside of the mitochondrion and in the open space of the cell's cytoplasm (Figure 1.13).

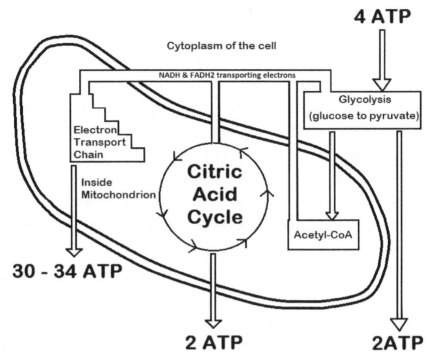

4 ATP

Cytoplasm of the cell

NADH & FADH2 transporting electrons

Glycolysis
(glucose to pyruvate)

Electron
Transport
Chain

**Citric
Acid
Cycle**

Inside
Mitochondrion

Acetyl-CoA

30 - 34 ATP

2 ATP

2ATP

Figure 1.13: Cell Respiration Overview

Glycolysis also breaks the glucose molecule into two smaller molecules. When those enter the mitochondrion the second step of cell respiration occurs. It is called the citric-acid cycle, or Kreb's cycle, after the German scientist who discovered it in the late 1930s. The remainder of the glucose is disassembled and then when the electons are moved through many steps of the electron transport chain energy is released, resulting in the production of 30 to 34 more ATP. The cell can make up to 38 ATP molecules from one glucose molecule. Because the cell uses two ATP molecules to jump start the process, the net gain is up to 36 ATP. Remember, this entire process is referred to as aerobic respiration because oxygen is required—hence the reason you need to breathe!

Video: Cell
Respiration
*https://vimeo.
com/167728269*

Let's return to the imagery of walking down that city street. You spot the bus stop two blocks ahead. And then breathe in the smell of diesel and feel the breeze created by the passing bus. That's your bus! Run! You begin to sprint. The bus begins to pull away but stops when the driver sees you sprinting to the bus stop. Phew! Panting, you climb on board, deposit your money and take a seat. Depending on your athletic conditioning, your body will run short of oxygen at some point even though your breathing rate has increased. When your cells run out of oxygen, they switch to a process of burning glucose without oxygen called *anaerobic* respiration. Adding the prefix "a-" (or "an-" in this case) to a term is like adding a negative sign to a number; it reverses the meaning. In your cells and in the cells of some other organisms, like a bacterium called *Lactobacillus acidophilus* (Figure 1.14), the process is referred to as lactate fermentation. Just as in aerobic respiration, the cell first breaks the glucose apart during glycolysis, resulting in four ATP. It then breaks down the rest of the glucose in the mitochondrion. But instead of transferring the electrons and capturing that energy to result in more ATP, the electrons are simply recycled and the additional ATP are not produced. Therefore, it takes much more glucose to power your body when you run out of oxygen than when you have caught your breath again and

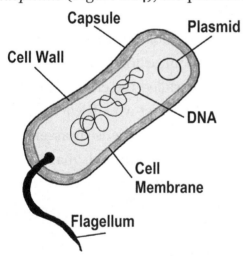

Figure 1.14: Typical Bacterium Cell

have plenty of oxygen flowing into your lungs, bloodstream, and eventually into the powerhouse mitochondria contained in your cells. Because each glucose molecule burned only nets two ATP (instead of up to 36 when oxygen is available), anaerobic exercise causes you to lose weight. As you exercise and force your cells to burn glucose without oxygen, your body burns off considerably more glucose than it does when resting. You first exhaust the glucose in your bloodstream from your most recent meals, and then begin to burn the stored energy reserves in your body, otherwise known as fat. The waste product of this anaerobic process is lactate or lactic acid.

The bacterium, *Lactobacillus acidophilus,* can only burn glucose when oxygen is absent, or anaerobically. When this bacterium is added to milk, it consumes the sugar in the milk through lactate fermentation, just as your cells do when you run short of oxygen. The build-up of lactate as the bacteria consume the sugar changes the milk into other dairy products such as buttermilk, cheese, and yogurt. Unless sugar is added to these products after the bacteria have done their work, these foods are not sweet because much of the naturally-occurring sugar in the milk has been consumed by the bacteria.

Some fungi also burn glucose without oxygen. Yeast, *Saccharomyces cerevisae,* is the most widely used. As in lactate fermentation, fewer ATP are produced than in aerobic respiration, but with yeast the waste product is ethanol (an alcohol) instead of lactate. This process is referred to as alcoholic fermentation. When yeast is mixed with water and a food source such as fruit, wheat, barley or hops, it consumes the sugars in the food source and the waste products are CO_2 and ethanol. This results in a carbonated, alcoholic beverage. Making bread utilizes this same process. Add yeast to flour, sugar, and water in a warm environment and the yeast will eat the sugar. The CO_2 gas that is released forms the bubbles that make the bread rise and the ethanol evaporates into the atmosphere, which is why you don't have to be 21 to eat bread!

These processes are happening in all living things. The plants use the food they make. You also use the food the plants make. You eat the plants (or what ate the plants), and store that food in your body as carbohydrates or fat. Your body burns the food just as the engine in a car burns gasoline. However, in a car, energy from the combustion of fuel moves the pistons in the engine, which causes the wheels to move. What does the "work" of causing things to "move" in the cells and in your body?

To understand this work, we have to return to a molecule previously mentioned: ATP. ATP is short for adenosine triphosphate (Figure 1.15), and is made up of an amino acid and a sugar, with a tail of three phosphate groups. A phosphate group is a phosphorous atom bonded with three oxygen atoms. This group plays an important role in transferring the energy

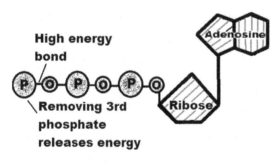

Figure 1.15: ATP Molecule

that started with the sun and eventually was stored in the form of sugars, carbohydrates, and fats. This phosphate group ultimately results in actual work being done within a cell. As these stored energy sources are broken down in cellular respiration, ATP is formed by adding the third phosphate onto an existing molecule called adenosine diphosphate. It takes a lot of energy to form this bond between the second and third phosphate group. It is this bond that is the last storage place of the energy that originated with the sun, which was first converted to chemical energy during photosynthesis, and will then eventually cause work to happen in a cell.

To see how this phosphate group makes work happen, let's return to our urban travels. The bus slows to a stop and you see your destination—a park with green space, walking paths, and playground equipment. Exiting the bus, you head towards the oasis of green in the middle of this concreted landscape. Notice the playground equipment with children buzzing around the slide, swings, and climbing apparatus. A good metaphor for how the cell stores energy is the young child at the base of a ladder leading to the top of the slide. To climb the ladder she has to use energy, and in so doing, creates the potential for releasing

Video: ATP, ADP and Energy
https://vimeo. com/167557705

energy as heat once she slides down. Friction between her rear end and backs of her legs and the surface of the slide causes the heat. Climbing the ladder is similar to the cell using the sun's energy to move the electrons away from the nucleus of atoms during photosynthesis and then making the glucose as one form of chemical energy. Cellular respiration completes the energy transfer and storage by breaking the glucose apart and using the energy released to add the third phosphate group onto the adenosine diphosphate, making yet another form of chemical energy in ATP. The action of letting go at the top of the slide and sliding down is similar to releasing the third phosphate from the ATP. When an organism needs work to happen inside the cell, it moves the ATP molecule near the molecule it needs to perform a function, such as a protein molecule in a muscle fiber. It took energy to add the third phosphate group to ATP. Removing that third phosphate from the ATP and moving it to the protein molecule in the muscle fiber uses this energy to change the shape of the muscle fiber, creating movement. So the transfer of the phosphate does cause actual, physical work; in this case it results in muscle movement.

Another way to think about this process is to compare it to the movement of a car. A raw energy source like oil is refined and put into the easy-to-use form of gasoline. This is similar to the cell taking the sun's light energy and converting it into glucose. The oil contains a lot of energy, but it is not directly used to cause work. Your car does not burn oil; it has to be converted to gasoline first. Just as with glucose, when the gasoline is broken down, it releases energy, which causes the pistons to move in the engine (like the formation of ATP in the cell during respiration), which causes the wheels to turn and move the car. The energy released when ATP is broken down causes muscle fibers to move, which causes your muscle to contract and move your body (Figure 1.16).

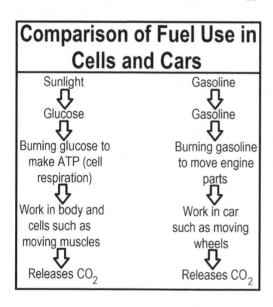

Comparison of Fuel Use in Cells and Cars

Sunlight	Gasoline
⇩	⇩
Glucose	Gasoline
⇩	⇩
Burning glucose to make ATP (cell respiration)	Burning gasoline to move engine parts
⇩	⇩
Work in body and cells such as moving muscles	Work in car such as moving wheels
⇩	⇩
Releases CO_2	Releases CO_2

Figure 1.16: Burning Fuels to Make Work

Understanding the flow of energy is key to understanding one's ecological identity. A good exercise to solidify your understanding of both the big picture and the details is to do the following. First, start with a flow chart like the one below, tracing the energy from the sun through the cell processes resulting in work by the cell. Then, create branches off each step of the flow chart with details, examples, and molecules used in each of those cellular processes (Figure 1.17).

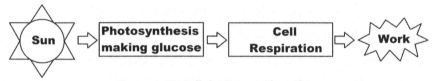

Figure 1.17: Cellular Energy Flow Chart

Energy flows through the ecosystem, mostly in a linear fashion. It enters one side from the sun, is captured by the plants, is used by all life, leaves in the form of heat back into the atmosphere, and eventually returns to outer space. Think of the earth as a balloon. If you continuously fill a balloon with air it will pop, unless you either stop blowing it up or it has a leak in it. The sun is constantly putting energy into the earth system. If the energy never left the earth, it would be like a balloon that was constantly being filled. The energy leaving has to equal the energy entering the system. But what is the medium that carries that energy?

5 – Recycling Matter

Back walking through our park, you have to step around puddles from last night's rain, and you wonder how "old" that water might be, recalling what you learned in school about the water cycle. Think about it this way: the water molecules in that puddle might have been in the ocean a few years ago, or even consider that when you showered this morning, you might have taken a shower in dinosaur spit! The hydrogen and oxygen atoms in the water molecules flowing from the showerhead have always been a part of the earth, getting recycled over and over and over again. The same is true about the oxygen and carbon atoms you breathe in and out (as a part of the energy system you have already read about). Those atoms have always been here. We have already seen how, through the burning of glucose created by the plants, you are an integral part of the flow of energy through the ecosystem. As you walk down that street, ride on a bus, and eat dinner, how are you, as a part of the ecosystem, involved in the *recycling* of matter on the earth? Matter is anything with substance or mass, anything made of atoms put together into molecules. How does this recycling of matter connect with your role in the flow of energy cycling *through* the ecosystem?

First, let's differentiate between energy and matter. Recall that energy can be transformed from one form to another, but cannot be created or destroyed. But where exactly is it? To understand this we have to first look at matter. You already know that you are made up of matter, or atoms. In living things, the most common atoms are carbon, nitrogen, hydrogen and oxygen, along with small amounts of phosphorous and sulfur. These molecules make up the bulk of your body and can be remembered using the acronym, CHNOPS, pronounced "shnops." Also, recall that energy is stored in the bonds between atoms, and that cell respiration is the process of breaking those bonds and capturing that energy. Your body, remember, stores that energy in the form of ATP molecules, specifically in bonds between the phosphate groups of the adenosine triphosphate molecule. If you put all of these atoms together, it actually weighs something. It has mass. When this mass is in a living thing, we call it biomass.

We have already seen the ecological connection between you and a plant in terms of energy. The plant converts the sunlight to chemical energy and stores it. You eat the plant and burn it. When eating the plant, you take the biomass synthesized during photosynthesis and use it to fuel your body. The plant got the carbon and oxygen in the glucose from the CO_2 in the air it absorbed through openings in its leaves called stomata (Figure 1.18). The hydrogen atom came from the water absorbed

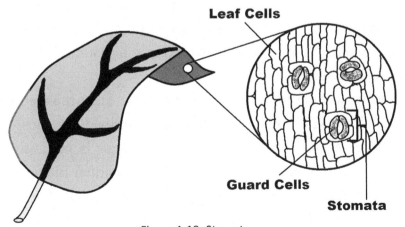

Figure 1.18: Stomata

through its roots in the ground. The O_2 released by the plant came from the water. Carbon makes up the majority of the mass of living things. It would seem to make sense that the majority of the mass of a plant comes from what it might suck up through its roots in the ground, but this is not accurate. The carbon, and therefore the majority of the atoms in a plant, all came from the air around you, *not* from the ground! It is easy to think of the air as being empty, but a majority of the biomass in all living things was once floating as CO_2 in the air. What happens to that carbon when you are done with it? You do not just *add* biomass. You must also lose biomass; otherwise, you would not be in a state of homeostasis. How do you lose it?

To answer this question we have to introduce decomposers to the story. Look around on the ground in the imaginary park. The ground is wet from the previous few days of rain. In the grass is a white mushroom. Decomposers are usually either bacteria or fungi, but we are going to focus on fungi for now. Fungi are the recyclers in the ecosystem. The plants build the molecules that make the biomass. The animals use those molecules and the energy stored within them to rearrange the atoms into yet more molecules and add to their biomass. Plants and animals are constantly adding and taking carbon and oxygen from the atmosphere, as well as water stored in the ground. But none of this could happen without the fungi.

Many varieties of fungi live on the planet and act as recyclers. Some are single-celled, like the yeast described previously, but most are multi-celled. The body of a fungus is made up of a mesh of filaments, or thin tubes. The individual filaments are called hyphae (singular, hypha). The term mycelium refers to the entire mesh of filaments put together. Fungi always have to grow on a food source. Most fungi have a mass of mycelia that grows into the food source (dead log, piece of bread, soil, etc.), and also produce some kind of reproductive part (most typically a mushroom) that is visible above the surface (Figure 1.19).

Common examples of fungi are mushrooms that we eat and mold that grows on spoiled food. That should be the clue as to how fungi fit into this system of recycling matter: a fungus grows on dead things. When we say they are decomposers, it is simply a way of saying fungi eat dead things. Though lacking a digestive system, a fungus can still eat its food. The fungus grows on a piece of fruit, dead animal, or log and sends its hyphae *into* the food source, se-

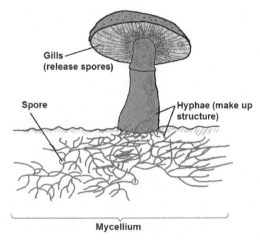

Figure 1.19: Typical Fungus Structure

creting chemicals and breaking it down and digesting it, thus extracting sugars out of the food, and then burning them with cell respiration. Mold growing on a piece of bread is doing the same thing that your digestive system does—just more slowly.

Scientists classify and put fungi into categories by how they reproduce. Sac fungi (phylum Ascomycota) are a type of fungus that includes yeast, many common blue-green molds, morels (Figure 1.20) and truffles (not the chocolate kind). They are named sac fungi because they produce a little sac that contains spores, which are used like seeds to spread the offspring away from the parent.

Another kind of fungus are the club fungi (phylum Basidiomycota), which have spores that are produced on microscopic, club-like structures called basidia. You might have seen examples of these called puffballs, which typically grow in wooded areas. If you step on one of these brown balls, it breaks open and the spores look like smoke as they float away in the wind. Toadstools are also club fungi. One of the most poisonous is *Amanita,* called the "death cap" (Figure 1.21).

Figure 1.20: Morel Mushroom

There is a group of fungi called "imperfect" fungi. Not because there is something wrong with them, but because their reproductive process is not readily understood. There are many individual species of fungi that are in the genus Penicillium, and many of them produce a chemical that is toxic to bacteria. This is a naturally-occurring defense mechanism in these fungi to protect themselves from bacterial infection. In 1928 Alexander Fleming was studying bacteria and discovered that a petri dish of a bacterial culture had been

Figure 1.21: Toadstool

contaminated with Penicillium. Bacteria was not able to grow in the areas directly around the growth of the Penicillium fungi. After further studies, he concluded that the mold secreted an antibiotic chemical that he named penicillin.

No matter the kind of fungus, its role as a recycler is the same. As fungi consume the food they are eating (or decomposing), they release CO_2 into the atmosphere just like you do when you breathe. But they also do more. They break down the organic molecules into smaller molecules. These molecules are then often broken down further by bacteria and into single atoms such as nitrogen, potassium and phosphorous, which are then returned to the soil. This is crucial because these nutrients are now in the soil and available for the plants to absorb along with water through their roots. Notice that carbon and oxygen are not returned to the soil, but instead released to the atmosphere when the decomposers do cell respiration to consume any sugars still left in the dead things they eat. The essential nutrient molecules that are returned to the soil do not provide energy to the plants, but are key ingredients in plant molecules and cells. These nutrients for the plants are like vitamins in your diet. They are not energy, but they are essential. That is why the key ingredients in plant fertilizer are nitrogen, potassium, and phosphorous. These are often called plant "food," but this is a misnomer because, as you already know, plants make their own food through photosynthesis.

Maybe the most important role that fungi play in the ecosystem is one that does not have anything to do with the recycling of matter through decomposition. Within the soil, the hyphae of fungi grow and spread. Fungal hyphae fill much of the soil, growing on and around the roots of plants. They steal some of the chemical energy made during photosynthesis that is stored in the plant roots as starch. The plants take advantage of the fungi as well. The fungi act like a sponge, creating little air pockets around the roots of the plants. When water runs into the ground—say from a rainstorm—instead of flowing straight through the soil, it gets trapped in the little air pockets among the mass of hyphae, making it easy for the plants to absorb.

6 – The Carbon Cycle

One's ecological identity is not just about how the environment influences an individual and species, but also the reverse. How the individual (and species) might change the environment. The most abundant atom in all living things is carbon. With the discovery that the earth's temperature is changing, due in part to the increased amount of CO_2 added to the atmosphere by human action, it is essential to understand how carbon and other atoms are recycled in the earth system through the actions of producers, consumers, and decomposers (Figure 1.22).

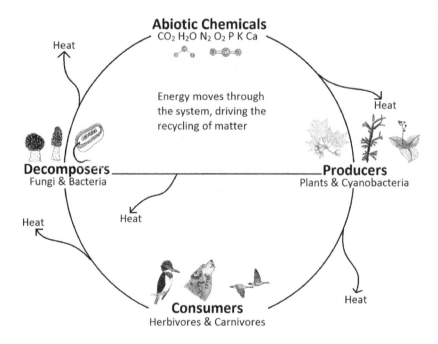

Abiotic Chemicals
CO_2 H_2O N_2 O_2 P K Ca

Energy moves through
the system, driving the
recycling of matter

Heat

Heat

Heat

Decomposers
Fungi & Bacteria

Producers
Plants & Cyanobacteria

Heat

Heat

Heat

Consumers
Herbivores & Carnivores

Figure 1.22: Matter and Energy Movement

In the atmosphere, a great deal of carbon is stored in CO_2 and methane (CH_4) molecules. We are going to focus on CO_2 because this molecule is directly connected to cellular respiration and photosynthesis. There are two processes of recycling carbon within the earth system. The first includes all the processes you have already read about that exchange CO_2 through photosynthesis and cell respiration. These processes move carbon between organisms and the atmosphere when one organism eats another organism, such as a cow eating grass, followed by you eating a hamburger. This system is in equilibrium because the amount taken in by one process (photosynthesis) equals the amount burned by living things (cell respiration). They tend to balance each other out. Because this all happens in the biosphere and within the atmosphere, scientists call it the atmospheric carbon cycle, or the biologic carbon cycle.

The other process of recycling the carbon in the biosphere is geologic. Imagine that a plant died in a swamp 150 million years ago. If the conditions were right, it was covered and compressed *before* fungi could decompose it, so the carbon in the plant was stored instead of returned to the atmosphere. After millions of years of being compressed under rock, it was physically transformed into oil or coal. Remember, energy can be transformed but not created or destroyed, so the energy is still in the molecules making up the oil or coal and now called fossil fuels. Fossil fuels are the product of millions of years of geologic processes. This fossil-fuel carbon is stored in what we call a "carbon sink," since it is stored underground away from the atmosphere and removed from the atmospheric cycle. Now imagine a power plant burning that

Video: Biologic vs. Geologic Carbon Cycle and Humans *https://vimeo. com/167562238*

coal. Recall that burning something is chemically very similar to your cells doing cellular respiration. Remember the waste product? As the coal is burned, the carbon in the coal is released back into the atmosphere in the form of CO_2 gas. The carbon and oxygen is released into the atmosphere through burning (respiration) but at a much faster rate than the millions of years it took to remove it from the atmosphere through photosynthesis and fossilization. CO_2 enters the atmosphere faster than the plants can absorb it through photosynthesis and the excess builds up. Much of that CO_2 is actually absorbed into the oceans. The oceans can also act as a carbon sink to store the carbon. Unfortunately, the more CO_2 accumulates in the oceans, the more acidic the oceans become, causing many species, such as coral reefs, to die.

We began this chapter with the image of walking down a busy city street, ending up in a city park and thinking about all of the connections to the life around us and our ecosystem. The evolution of our species and society is a story we all participate in creating with each breath we take, the food we eat, and the energy we use. That story can actually be traced back hundreds of thousands, even millions of years. Imagine one of our human ancestors, a fellow member of our genus, a human-like creature called *Homo erectus* (Figure 1.23), using fire for the first time one million years ago. Initially, fire was used as a source of heat and to protect against nighttime predators, and the smoke possibly warded off biting insects. This might be the first example of a member of our genus using a fuel

Figure 1.23: *Homo erectus*

source indirectly instead of ingesting it. Fire usage by *Homo erectus,* and then by early *Homo sapiens,* expanded to hardening tools and cooking food. Cooking would have made food easier to digest and allowed individuals to have a broader choice of foods in their diets. This use of fire to prepare the foods we ate was a shift in our ecological identity, as it caused a shift in how we interacted with the environment and how the environment then influenced our biological and cultural evolution.

Our ecological identity is now greatly influenced by the use of fossil fuels. It is in changing the geologic carbon cycle where humans have had the greatest impact. Coal use began as long ago as 2,000 years ago in the Roman Empire. However, it was not until the industrial revolution in the eighteenth and nineteenth centuries that coal use really ramped up. The first coal mine in the United States opened in the 1740s. It wasn't until the 1880s that coal was used to produce electricity. Still today, a great majority of the electricity in the United States comes from the burning of coal.

Like coal, crude oil has about a 2,000-year history of use. However, initially crude oil was not used as a fuel source, but as a lubricant, as part of religious rituals, and even as a medicinal bath. There is evidence that 2,000 years ago oil was used in limited quantities as a fuel source to boil brine water to make salt. As long ago as the tenth and twelfth centuries, oil was a fuel source in lamps. Extensive burning of fossil fuels did not begin to occur until the advent of motorized transportation and machinery at the beginning of the twentieth century, however. Crude oil is not burned directly in our cars, but instead it is the gasoline and diesel fuel refined from the crude oil that is burned. Producing gasoline and diesel takes a great deal of heat, so even the process of refining the oil requires fossil fuels.

Directly and indirectly, you as an individual, and all of us collectively as a society, are key to the movement of carbon, oxygen, and hydrogen throughout the earth system. You consume molecules and break them down, releasing energy. You release waste products in the form of gases, solids, and liquids, and eventually contribute your decomposed body after your death. It is all recycled. Recycling this matter requires energy to drive the system and keep it functioning and in homeostasis. That is the role of sunlight. Much like how a living cell or human being uses energy to organize matter into an organized structure to create the miracle of life, the earth too, as a system, uses the sun's energy to maintain a relatively constant temperature—thus allowing us to live comfortably within the biosphere. We can return to our metaphor of the car using energy to better understand this. As a living system goes down "the road of life" maintaining homeostasis, it uses fuel to do so, and gives off waste and heat just as the car gives off exhaust and heat when it drives down the road. Energy passes through the system and is the fuel to maintain the complex living systems, making order from chaos.

Indirectly, then, any actions you take that might change the homeostasis of the planet by changing the rate of the recycling of the key ingredients of living systems and the earth system could be changing the planet's ability to maintain homeostasis, and ultimately affecting our ability to survive within that system.

Exploring Biodiversity Web Page
www.exploringbiodiversity.com

Chapter One Web Resources *http://www. exploringbiodiversity.com/#!blank/c1tc5*

Chapter 2: Ecosystems

How are you a part of a dynamic ecosystem?

1 - Nature's Recyclers

Picture the scene of our city park filled with mature red oak trees, *Quercus rubra,* providing shady spaces throughout the green lawn. There are large boulders and rocks here and there, as well as flower beds and bushes. Birds, insects, and small mammals such as squirrels move about. And of course you are here too.

Beyond the obvious pleasure of running around barefoot on grass in the shade, there may be a biological reason that we humans tend to build parks with open grassy areas dotted with shade trees. Evidence indicates that as long as 3 to 4 million years ago, one of our human ancestors, *Australopithecus afarensis*, made the transition from living as a woodland creature to inhabiting the grasslands and open savannas (Figure 2.1). It could be that the adaptation of upright walking gave this species an advantage over other, less upright walkers by allowing them to see oncoming predators and find food sources. There were even more ancient members of our family tree that walked upright, but the evidence is less clear as to what type of

Figure 2.1: *Australopithecus afarensis*

habitat they preferred. In any event, we have millions of years of living in open spaces as a part of our genetic and evolutionary history as a prime component of our ecological identity.

We will continue to come back to your role in the ecosystem, but it is not always about you! There are countless other species on the planet—from bacteria and algae, to giant redwood trees and whales. And they all play important and specific roles in a dynamic ecosystem. They all help an ecosystem change, adapt to changes, self-regulate, and maintain homeostasis.

Let's begin with the smallest, yet most abundant of organisms in the ecosystem: prokaryotes. These are simple, one-celled organisms without many cell parts. The name prokaryote refers to this simple cell structure. They are commonly called bacteria, but bacteria are actually just one of two kinds of prokaryote. Prokaryotes make up two of the three domains of organisms on the planet: Archaea, Bacteria, and Eukaryota. Bacteria and Archaea are prokaryotes. All other life on the planet are eukaryotes, meaning they consist of cells with complex parts inside of them called organelles. The "typical" plant and animal cells described in Chapter One fit into this category. In addition to plants and animals are fungi and protozoans. It is unknown how many species of prokaryotes live in the biosphere—possibly millions of different species. Without a doubt, a great diversity of prokaryotes fills a variety of niches, or roles, in the biosphere. Some variety of prokaryote covers everything—probably more than you can imagine.

Return to the park and imagine children running around in the play area, all touching the same pieces of equipment. Consider the germs being passed from one child to the next from hand to playground equipment to the next child's hand. If you look closely at the palm of your own hand, it appears smooth and soft. To microscopic prokaryotes, however, your hand is a mountainous ecosystem. On the skin's surface and tucked into tiny crevices are a wide variety of prokaryotes. Astonishingly, there might be more prokaryotic cells on and in your body than actual human cells, up to ten times more according to recent research. However, there is no need to panic. An alien force has not invaded and overtaken your body. This is how it should be. You live in symbiosis, or in collaboration, with these organisms. This is just one of many interactions you have with another organism as a part of your ecological identity.

Sitting on the bench near the play area of our park is a mother breast-feeding an infant tucked under a blanket draped over her shoulder. This is where the symbiotic relationship between you and bacteria may have begun. A baby ingests prokaryotes when he or she breast feeds and also simply by being nestled up to his or her mother and in contact with her skin. Scientists estimate that up to 500 different species of prokaryotes live in the typical human intestine. They help you, and you help them. Prokaryotes aid in digestion and help to keep your immune system healthy. This relationship is ancient, indicated by the fact that approximately 40 of the 20,000 genes in the human genome have a prokaryotic origin. If these prokaryotes do not help you, they at least do not hurt you, and simply are hitching a free ride. Of course, there are some that do cause illness. Harmful prokaryotes cause symptoms of illness when your body identifies them as foreign invaders and fights them off by calling on the white blood cells of your immune system to engulf and destroy the smaller bacteria cells (Figure 2.2). One symptom is an increase in your body's temperature, causing you to have a fever.

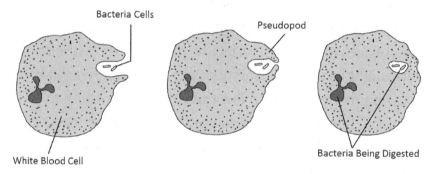

Figure 2.2: White Blood Cells Engulfing Bacteria

The simple-celled, bacteria-like prokaryotes in the group Archaea are possibly very similar to the first living things on the planet, dating back as far as 3.5 billion years. Modern examples of Archaea tend to be found in extreme places on the planet, like deep-sea volcanic vents. Therefore, it is doubtful that you have this kind of prokaryote living on your hand.

The more common prokaryotic organism we regularly have contact with is from the domain Bacteria. Primitive species from this group were likely the first organisms to develop photosynthesis. This adaptation of photosynthesis had a major impact on the atmosphere, since, as you already know, the waste product of photosynthesis is oxygen. Conventional wisdom in the scientific community has been that the earth's early atmosphere contained very little oxygen, high levels of methane, ammonia, carbon dioxide, and water, and some molecules containing sulfur. Recent evidence from researchers at Rensselaer Polytechnic Institute indicates that the Earth contained higher levels of oxygen than previously thought. This is an example of the always changing and tentative nature of science. Despite uncertainty about the precise makeup of the early earth atmosphere, scientists agree that ancient cyanobacteria first evolved the ability to photosynthesize sugars using energy from sunlight and thus began adding oxygen to the atmosphere. One of these more common prokaryotes in the bacteria domain—though not a photosynthetic species—is undoubtedly the kind of bacteria living on your skin.

Species in the bacteria domain are categorized based on the shape of their single cells. Bacteria similar in structure to a rod or cylinder are called bacillus. Spiral or corkscrew-shaped bacteria are called spirilla. Cocci are spherical-shaped, like a ball. The bacterium that causes strep throat is this form and is known as Streptococcus (Figure 2.3).

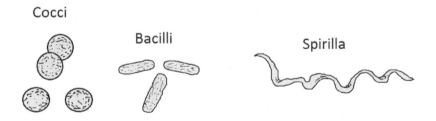

Figure 2.3: Three Bacteria Shapes

No matter the shape, all these bacteria have some common characteristics. They are single-celled but can live in colonies or clumps. They are *very* simple cells. They have a cell membrane and a cell wall, but have no complex parts inside, unlike your cells or the cells of all the other groups of living things on the planet, the eukaryotes. They do not even have a nucleus with the DNA stored safely inside, but they do have DNA made up of the same chemical components as your DNA. This commonality is some of the evidence that all life on the planet evolved from a common ancestor.

Video: Bacteria
Shapes and Sex
*https://vimeo.
com/167567611*

Instead of storing DNA inside of a nucleus, like the eukaryotic organisms, bacteria have a single loop of DNA called a nucleoid. The nucleoid contains the instructions for what happens in the cell. The cell has another form of DNA, called plasmids. These are smaller circles of DNA that are transferred readily from one bacterium cell to another cell. Bacteria do not have sexual reproduction like more complex eukaryotes, but they can transfer plasmids with other bacteria as a substitute for sexual reproduction. Interestingly, it is through our understanding of how to manipulate these plasmids of DNA that humans have been able to use bacteria plasmids as vectors (or messengers of a sort) to insert genes from one organism into another. This is the technology that has allowed us to begin doing genetic engineering, which will be explored further in Chapter Five.

After all that background information, let's return to the topic at hand—bacteria's role in maintaining homeostasis of the biosphere. Like fungi, bacteria are important decomposers, feeding on dead organisms and extracting energy from the organic molecules. Remember, this is no different from when you burn energy stored in the sugar and carbohydrate molecules in the food you eat. Bacteria do not use a digestive system as you do, but instead operate like fungi in obtaining sustenance. It is as if their stomachs were on the outside. Bacteria do more than just consume other organisms. Think of them as super-digesters. Consumers, like you, break down food to get the energy out, but they only break it down so far, leaving much of the waste as large molecules. Decomposers go a step further and break the molecules down into individual atoms, or at least into smaller molecules, releasing elements like phosphorous and nitrogen into the ecosystem (Figure 2.4).

Not all bacteria are decomposers. Some bacteria use photosynthesis, much the same way plants do. These are the cyanobacteria, sometimes called the blue-green algae even

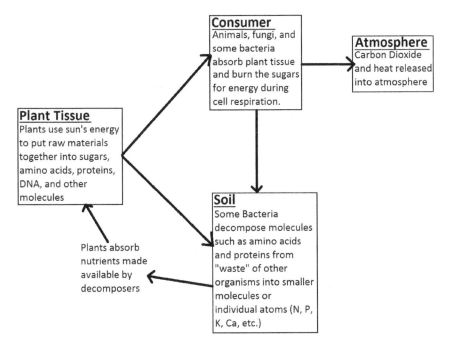

Figure 2.4: Consumers vs. Decomposers Flowchart

though the term algae is usually reserved for labeling plants. The cyanobacteria live in water habitats, from hot springs to Antarctic lakes, and are responsible for algal blooms during warm summer months. This is when there is a tremendous growth of algae in a lake, making the water green and even kind of soupy. Some bacteria can feed on inorganic molecules instead of molecules from once-living things. These are called chemoautotrophs. They are referred to as chemosynthetic because they use molecules from the environment to make (or synthesize) their own food, much like photosynthesis, but instead of harnessing the sun's energy to do biosynthesis, they use molecules found in the environment. Both chemosynthetic and decomposer eubacteria are responsible for a crucial biogeochemical cycle that helps the biosphere maintain homeostasis. This crucial process is the nitrogen cycle.

2 – The Nitrogen Cycle

Let's continue our walk through the park and proceed down a path away from the playground area. As the sun climbs higher in the sky the air you breathe in is warmer. The majority of that air is actually nitrogen, not oxygen. Recall that nitrogen is a crucial atom in all living things. It is part of the acronym CHNOPS, along with carbon, hydrogen, oxygen, phosphorous and sulfur. Nitrogen is a key ingredient in amino acids, which are the molecules that your body uses as the building blocks for all of its protein molecules. Nitrogen is also a key atom in the DNA and RNA molecules inside your cells' nuclei.

Nitrogen is extremely abundant in the atmosphere. The majority of the atmosphere is made up of nitrogen in its stable gaseous form, which is two nitrogen atoms bonded together (N_2). Unfortunately, neither plants nor animals can absorb it in this form (Figure 2.5). But because it is essential to all organisms' survival, they must get it somehow.

Nitrogen requires modification before it can be absorbed by plants. We call this "nitrogen fixing." A great deal of energy is required to fix the nitrogen, which simply means breaking apart the two nitrogen atoms and causing them to bond instead with available hydrogen or oxygen atoms. The lightning during thunderstorms can cause this process to happen spontaneously. After the nitrogen

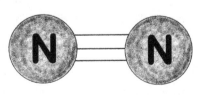

Figure 2.5: Nitrogen Molecule

atoms are broken apart, they bond with available oxygen in the atmosphere to form nitrate ions (NO_3^-). Plants can directly absorb this molecule, unlike the N_2, and then incorporate the nitrogen into their proteins. Unfortunately, naturally-occurring lightning does not provide nearly enough usable nitrogen for all the world's plants. This is where bacteria enter the story.

Bacteria fix the majority of nitrogen used by plants. For example, one type of nitrogen-fixing bacterium lives in little bumps called nodules on the roots of legume plants (Figure 2.6)

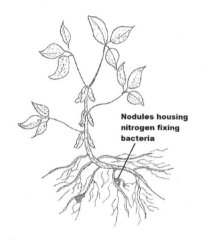

Nodules housing nitrogen fixing bacteria

or can live freely as independent-living organisms in the soil. Legumes are bean plants, like soybeans, peas or green beans. Legumes and bacteria live together. Bacteria provide the legume plants with fixed nitrogen and legumes provide the bacteria with food energy made during photosynthesis and stored in the plant's roots.

Figure 2.6: Legume Plant With Nodules

This is an example of a symbiotic relationship called mutualism.

Nitrogen-fixing bacteria break apart the N_2 molecules from the atmosphere and combine them with hydrogen atoms, forming ammonia molecules (NH_3), which are then converted into ammonium (NH_4^+) and nitrates (NO_3^-). These two molecules are readily dissolved in water, allowing the plants to absorb the nitrates when they draw water out of the soil. The plant cannot absorb the ammonium (NH_4^+) or ammonia (NH_3). However, other species of bacteria convert the ammonium (NH_4^+) into nitrite (NO_2^-), which in turn is absorbed by yet another bacteria species, which then converts the nitrite (NO$_2^-$) into nitrate (NO$_3^-$), which the plants can absorb. It is a complex cycle, but the takeaway is that plants rely on bacteria to convert nitrogen gas into an absorbable form. Without nitrogen atoms, plants could not produce DNA or protein molecules as they grow.

Nitrogen can be fixed through a human-made process called industrial fixation. Atmospheric N$_2$ is cooked under high heat in a factory and is converted into ammonia and nitrates. The end product is chemical fertilizer for lawns, gardens, and farm crops. Fertilizer is used to add a great deal of usable nitrogen to the soil for plants to absorb directly. It is probable that the grass in this park has been fertilized with a chemical fertilizer to add usable nitrogen to the soil. The additional ammonia is eventually converted by soil bacteria into nitrates as well. Fertilizer is especially necessary in areas when harvesting the crops involves physically removing the plant material from the location. Those plants absorbed a lot of nitrogen from the soil. By removing the plant material, you are also removing the nitrogen and removing the plant material so that it cannot be decomposed back into that soil. Those nutrients are permanently removed from that specific spot. Growing legumes that house the nitrogen-fixing bacteria in nodules on their roots or adding chemical forms of nitrogen can replace the lost nitrogen in the soil. Using chemical fertilizer comes at a cost, however, as is always the case when a naturally-occurring process is sped up, slowed down, or stopped altogether, and the homeostasis of an ecosystem is affected. Interrupting these natural cycles is an all-too-common component of our species' ecological identity.

Video: Consumers, Decomposers and Nitrogen
https://vimeo.com/167569139

Understanding the consequences of interrupting that system requires completing our description of the nitrogen cycle. So far, I have only described how nitrogen gets into the system from the atmosphere. Plants absorb nitrogen after it is fixed. Now it is in the living, or "biotic" part of the ecosystem. As stated before, the plants incorporate nitrogen into their amino acids and DNA. When an animal eats a plant, it gets those molecules and the nitrogen along with it. The animal then incorporates those molecules and DNA parts into their own cells. The molecules that make up DNA and proteins in all living things are interchangeable, and therefore animals can use plant protein and DNA pieces in its own cells and molecules.

Nitrogen leaves the animal in one of two ways: as waste or when the organism dies. In either case, decomposers break down the waste or the animal's dead body, combining the nitrogen with hydrogen to make ammonia, which can then be nitrified into nitrites and nitrates by bacteria and recycled yet again. This creates a cycle within a cycle (Figure 2.7) and effectively keeps the nitrogen in the biotic part of the ecosystem in that specific location.

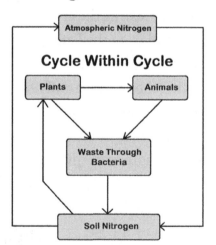

Figure 2.7: Atmospheric & Biologic Nitrogen Cycle

But how does it get back into the atmosphere? Clearly it does somehow because the amount of nitrogen in the atmosphere is a constant 78%.

Nitrogen leaves the ecosystem in three ways. One involves yet another kind of bacteria doing the process of denitrification. Just like the name implies, this is the *opposite* of nitrification. Remember that during nitrification bacteria convert N_2 first into nitrites and then into nitrates. Denitrifying bacteria use nitrates for energy and convert the nitrogen back into its stable gas form, N_2. These molecules then float from the soil and into the air.

Excess nitrogen also leaves the system in nitrate form when it is washed away as a pollutant into lakes, streams, and groundwater. This is the cost of chemical fertilizers. Any extra nitrogen from applied fertilizer ends up in ground or surface water. The cyanobacteria and algae living in that water now have an easy nitrogen source. Recall that cells need nitrogen to make new amino acids (protein components) and DNA. Increasing the amount of nitrogen available to these organisms in aquatic ecosystems is one factor that allows them to grow rapidly and increase their population beyond normal levels. When a bloom of algae and small photosynthetic plants grow and reproduce quickly it is called an algal bloom. One would think that all of these photosynthetic cyanobacteria and algae growing in a lake would increase the amount of oxygen in the lake and that this would be good for the ecosystem. However, the cyanobacteria and algae also die quickly and greatly increase the amount of dead, decaying matter in the ecosystem. When the bacteria, acting as decomposers, do cell respiration to get the energy from this food source, they use up the free oxygen in the water and increase the carbon dioxide content. This increases the acidity of the water as the oxygen levels decrease, causing other animal life in that system to suffocate.

Another way nitrogen leaves the system is simply by people harvesting the plants. Think of it this way: if you grow corn in your garden and you cut down the corn, haul away the stalks and eat the kernels, the nitrogen that normally would have been returned to the system by a decomposer is now moved to a new location or even an entirely new ecosystem. If the nitrogen is not returned to that specific soil, the total amount of nitrogen in that location is decreased. To continue to grow plants there, the nitrogen has to be replaced by either growing legumes or adding fertilizer—either chemical or organic (usually composted plant material or manure from animals like cows, horses, or chickens). When left to its own devices, the biosphere utilizes the biotic and abiotic mechanisms to maintain a constant level of N_2. This is another example of the planet maintaining homeostasis, and is also an example of how human actions are directly connected to and affect the homeostasis of an ecosystem (Figure 2.8).

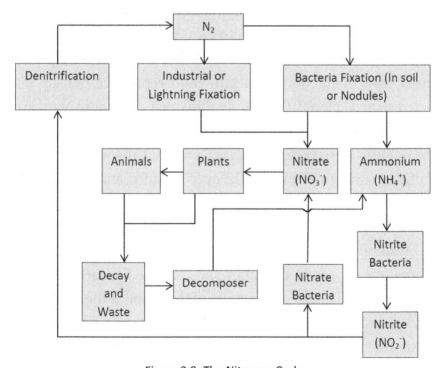

Figure 2.8: The Nitrogen Cycle

3 – Interacting Organisms

Let's return our focus from the microscopic world to the macroscopic world and broaden our vision beyond the bacteria living on your hand to other organisms around you in the park. What else might live here to make this ecosystem function, and what roles do they fill? What are their niches? The plants of course are photosynthesizing and storing chemical energy. These are the producers. Consumers are organisms that cannot make their own food. In addition to being consumers, the bacteria and fungi are also functioning as decomposers. All these organisms living in an ecosystem are linked together in what is called the biological community, for the shared purpose of maintaining the ecosystem. Recall that the ecosystem is all the living (biotic) and non-living (abiotic) things in the area, while the community is just the living things. All these living things are interacting. We have already explored many of these interactions, but let's look at some more.

Picture a shrub in the park. It stands ten to fifteen feet tall, with many stems and green leaves, and is full of purple berries. A red songbird, a Northern cardinal, *Cardinalis cardinalis* (Figure 2.9), moves between the branches, eating the berries. In this scenario, the plant is the producer and the bird is a primary con-

Figure 2.9: Northern Cardinal

sumer. The bird gets energy from the plant. Of all the energy the plant stored through photosynthesis, only about ten percent makes it to the bird. The plant uses the rest to power its growth and daily living needs, and eventually the energy not stored in sugars is lost to the atmosphere as heat (see Chapter 1).

Next, a least weasel, *Mustela nivalis,* might eat the cardinal (Figure 2.10). Though unfortunate for the bird, it is the weasel's day to have good fortune. The weasel in this relationship is a secondary consumer. Of all the energy the bird obtained by eating the plant, only ten percent of that energy actually

Figure 2.10: Least Weasek

gets to the weasel. The bird, like the plant, used the other ninety percent to live and grow, and that ninety percent eventually returns to the atmosphere as heat. Even though it may have been a good day for the weasel to get the bird, the weasel's luck is about to run out. A great-horned owl, *Bubo virginianus,* eats

Figure 2.11: Great-Horned Owl

the weasel (Figure 2.11). The owl is a tertiary consumer. It only gets ten percent of the energy that passed through the weasel, which only gets ten percent of the energy that passed through the cardinal, which only gets ten percent of the energy that passed through the plant. These are examples of predator-prey relationships. Those doing the eating are predators, and those being eaten are prey. Many organisms can be either one, depending on their fortunes on any given day.

Many of these organisms may die without being killed by a predator. When they die of "natural" causes, scavengers often eat them. One such scavenger is the turkey vulture, *Cathartes aura*. You have probably seen these birds circling overhead and thought they were an eagle or a hawk. They are large and soar for long periods of time without flapping their wings. You can tell turkey vultures from eagles and hawks because their wings angle up from the body, giving them a tell-tale "V" (for vulture) shape. They also rock back and forth in the air, kind of like a teeter-totter, when soaring.

Figure 2.12: Food Chain

These predator-prey relationships are organized into a food chain (Figure 2.12). If you connect all the food chains in an ecosystem, you get a food web. Mapping a food web can quickly get complicated, as most organisms eat more than one type of prey species or have multiple predators. This means that most organisms are involved in multiple food chains (Figure 2.13).

Returning to our discussion of energy transfer within the food chain, we can see how energy travels up the food chain. Ecologists use a pyramid to show the energy levels. The word "trophic" means energy, so it is called a trophic pyramid (Figure 2.14). Each step up the pyramid represents a different portion of the food chain. Using our example from the park, berries occupy the space at the bottom, followed by the bird, the weasel, and finally the owl.

Pyramids are largest on the bottom and most of the ecosystem's energy is stored here in the plants. This is where the sun's energy enters the system. The primary, secondary and tertiary consumers occupy each step up the pyramid respectively. Because ninety percent of the stored chemical energy is lost as heat with each step up the pyramid, it is easy to see why there are so many more plants and small animals than there are large animals at the top. It takes many plants to feed the primary consumers, many primary consumers to feed the secondary consumers, and so on.

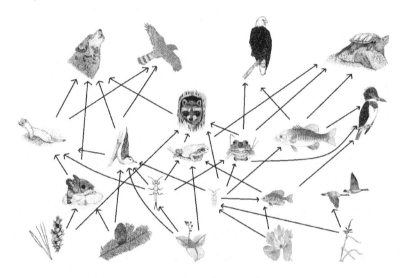

Figure 2.13: Food Web

Think about the movement of the energy through the food chain in terms of biomass from the first chapter. Recall the laws of thermodynamics: energy cannot be created or destroyed; also, energy can be converted from energy to mass and back to energy again. This is what the producers do. They convert the sunlight into the biomass of the plant cells. The plants form the base of the pyramid with the largest amount of biomass.

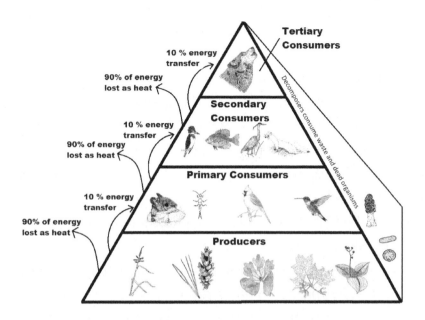

Figure 2.14: Trophic Pyramid

They get eaten, but most of the energy is used by the plant and escapes from the food web as heat. Since most of the energy does not get transferred, and the organisms at each step need the energy from the step below to organize the atoms and molecules into biomass, each step up the pyramid has to contain fewer organisms. Also, each organism up the food chain tends to be larger than those below, meaning each individual contains more mass, and since there is less energy available to convert into biomass, the organisms are fewer in number but larger. All organisms ultimately suffer the same fate, however. No matter where they are in the food chain, all are eventually consumed by decomposers, which utilize any energy left in their molecules and then break

Video: Predator-Prey and Energy Transfer
https://vimeo. com/167569659

the rest down into smaller molecules and atoms, returning them to the soil for the plants to absorb and begin the process all over again.

Predator-prey relationships are the easiest to see because the impact of one species on another is immediate, but of course there are other relationships within any community. Each species within an ecosystem has a specific niche, and no two species can fulfill exactly the same one. While each is important in the survival of the ecosystem, a species can be removed and have minimal effect *if* there are enough other *similar* species within that ecosystem. Sometimes, however, an ecosystem contains a species called a "keystone species." The removal of this species dramatically changes the ecosystem, because the keystone species directly or indirectly affects the survival rates of many other species.

Imagine yourself snowshoeing in Northern Minnesota. You come to a frozen pond with a stream feeding in one side and out the other side. As you snowshoe across the pond, you notice a five-foot tall, snow-covered mound protruding through the ice—a beaver lodge (Figure 2.15). Quietly, you walk up to the lodge, thump your snowshoe on the side and listen. First one

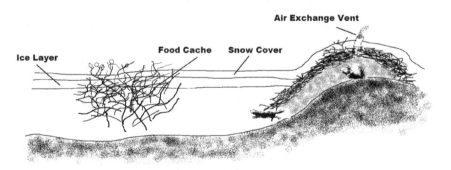

Figure 2.15: Beaver Lodge

splash, and then another, as the adults living inside exit a hole through the bottom of the lodge, escaping to safety into the water below the ice. Walking further down the pond, you discover that the pond exists only because a dam made by the beaver, *Castor canadensis,* has backed up the stream, flooding this low spot with just the right amount of water. This provides the beaver with a safe winter home to store its preferred food, the new growth at the top of trees. The beaver uses its oversized front teeth to cut down tall trees to get the tasty tops, which it drags into the deep water near its lodge, creating a cache of food for winter. The dammed-up stream or river freezes over, but the beaver can swim below the ice and get food all winter long. A beaver is a keystone species. By building dams to create deep water, they alter an area's habitat, changing the available niches for organisms to fill. By cutting down many trees, they also directly affect the tree population and indirectly affect the other organisms that are reliant on those trees.

The kangaroo rat, *Dipodomys deserti* (Figure 2.16), is also a keystone species. Researchers conducted a twelve-year study of kangaroo rats in the Arizona Chihuahuan Desert. When the kangaroo rats were removed from sectioned-off areas of the desert, the habitat changed from one of sparse grasses and shrubs to much more dense grassland within a few years. Researchers discovered that as the rats hopped around the desert, they kicked up seeds, exposing the seeds for birds to eat. Re-

Figure 2.16: Kangaroo Rat

moving the kangaroo rats allowed more seeds to germinate because now the birds couldn't find them. The grass plants spread and filled in more spaces, thus changing the habitat and consequently affecting the species that could live in that location.

A relationship between two organisms is called symbiosis. In common language, this term usually means that the organisms are helping one another. We should be more precise, however. Mutualism (a type of symbiosis) is the proper term defining a relationship where two species interact to the benefit of both. They mutually help one another. The nitrogen-fixing bacteria living in the nodule of the legume plant is an example of a mutualistic relationship.

Commensalism is another type of symbiosis, where one organism benefits from the relationship and the other is unaffected. Take yourself back to the park and imagine that a spider has built a web in one of the trees. The spider definitely benefits from the tree, because it now has a place to build a web. The tree, however, is neither harmed nor helped by the spider.

The third type of symbiosis is parasitism. This relationship involves one organism living on or in another. The parasite is benefitting and the host is being harmed. A fungus growing on the side of a tree is a parasite, as is a tapeworm growing inside the intestine of a dog. Often, the parasite does not kill the host (like a true predator does) but instead removes nutrients from the host while it lives, thus causing harm to the host, slowing its rate of growth and possibly shortening its lifespan. However, if the host does die, then the parasite will have to find a new host.

The fourth type of symbiosis is amensalism. Amensalism is similar to parasitism, because one organism is harmed, but the other organism is not directly helped by the relationship. In the case of amensalism, one species inhibits or obliterates the other one inadvertently. A sapling growing under a large tree is an example of this type of relationship. If the larger tree provides too much shade and absorbs most of the water, the sapling may not survive, but the larger tree is unaffected. For another example of amensalism, think back to the effect of adding too much fertilizer to a garden, lawn or farm. The excess runs off into a nearby lake and provides nutrients (nitrogen or phosphorous) for the cyanobacteria or algae, which causes an algal bloom. The actions of the humans spreading fertilizer helps the photosynthetic organisms, which then negatively affects the other organisms living in that lake. This is one more example of an aspect of our ecological identity in which humans might influence an ecosystem, and more specifically, affect the biodiversity of an ecosystem.

4 – Humans and Biodiversity

Biodiversity is the diversity of life within an ecosystem. The park you are imagining yourself walking through is most likely low in biodiversity. Visualize what you might typically see in a city park. It is dominated by grass—almost certainly a type bred and planted by humans—with a few species of trees but very little shrubbery or underbrush in order to keep the park looking "neat and clean." As far as animal life goes, there are small birds and mammals but very few predators, and certainly no large predators. Large predators such as coyotes, panthers, and wolves do not mix well with cities. Therefore, the ecosystem of this park is essentially incomplete, with the top portion of the trophic pyramid missing.

Is this lack of biodiversity a bad thing? "Bad" is a subjective term, not a scientific one. However, this ecosystem is certainly less stable than if it consisted of a wider variety of species. Think back to our food chain. Remove the weasel and if there is no other predator to help control the bird population or to provide food for the owl, that food chain becomes unstable. Imagine that in addition to the weasel as a secondary consumer, there are also other small predators such red fox, *Vulpes vulpes*, bull snakes, *Pituophis catenifer*, and raccoons, *Procyon lotor*. If a wide variety of species exists at every level of the trophic pyramid, then removing one of them (say through amensalism) might have less of an impact on the ecosystem's stability and homeostasis. The question we need to focus on is how human actions affect biodiversity.

E. O. Wilson is an ecologist who has spent his career at Harvard University. He created the acronym, HIPPO, to categorize the various impacts humans have on the biodiversity of the planet. The "H" stands for habitat destruction. Place yourself back in the park. What was here before it was a park where now it is grass, flower gardens, playgrounds, fountains, benches and walking paths? Picture the street and sidewalk we imagined ourselves walking on to get to this park. What was there before? To make homes, neighborhoods, and even green spaces for ourselves we destroy the habitats of other organisms. This is inevitable. Lumber for the homes requires removing trees. Clearing space for the homes requires removing most of the plants living there. Removing the plants affects the animals and vice versa, just like removing the kangaroo rat from sections of the desert changed it to grassland. To dig a foundation for our home, make roads, paths, playgrounds, and fountains requires digging up the soil and therefore disturbing the bacteria and fungi living in symbiosis with the plants that require the fungi to aid their water and nutrient absorption.

Habitat destruction doesn't end with the building of homes. Feeding ourselves requires farmland. Not only does farming usually require the removal of all existing life from an ecosystem in order to replace it with a single crop—the least possible biologically diverse ecosystem that can exist—the resulting lack of diversity requires continuous maintenance to keep it a monoculture of our preferred food choice. The majority of the electricity currently powering our homes comes from coal, the mining of which often requires removing entire mountain tops and all that lives in those ecosystems. These are just a few, obvious examples of the type of habitat loss that is the primary reason for a loss of biodiversity. Knowing that greater levels of biodiversity create more stable ecosystems, are you willing to change your lifestyle to reduce the need for these resources? Can you live with a smaller home, with less electricity, with fewer food choices? In addition to these obvious examples of human impact, there are more subtle forms of amensalism at the hands of human activity. As you read the next few paragraphs, see if you can identify examples of amensalism, and continue to ask yourself the difficult question: what would I be willing to give up—if anything?

The "I" in HIPPO stands for invasive species. In our imaginary park, the grass that makes up the lawn is probably not native to that ecosystem, but instead was transferred into the park in order to make a nice lawn. It is common for people to transplant species from one ecosystem to another. We like to manage the space around our homes, so we plant species that we find pleasing to the eye, which produce fruits and vegetables we like to eat, or that attract the animals we like to watch, such as butterflies, birds, or deer. This is part of our current ecological identity. We like to control the natural world with which we have immediate contact.

Oftentimes, however, the organisms we bring to a new location are called "exotics." Exotics are simply species transported to a habitat other than the one in which they evolved. Many times, these exotics become what ecologists call invasive species. A transplanted organism becomes invasive if—because it did not evolve in that ecosystem—it has no predators or it has some advantage over the native organisms, and then is able to dominate and choke out the native species. Not all invasive species are brought to an ecosystem purposefully. An invasive species can also "hitchhike" in boxes or packaging from one part of the world to another, and then enter a new ecosystem inadvertently. This is common with small animals like insects or other invertebrates (animals without internal skeletons).

Some years ago, as I snorkeled in Lake Michigan, I was amazed that as soon as I put on my mask and dipped my face into the water, I could clearly see the bottom of the lake. I went out ten feet, then twenty feet, then thirty feet, and I could still easily see every rock and pebble at the bottom. But mostly what I saw was thousands of small, filter-feeding mollusks called zebra mussels, *Dreissena polymorpha* (Figure 2.17). This species has invaded the Great Lakes of North America, as well as many of the smaller inland lakes. Cargo ships from other parts of the world unknowingly transported the zebra mussel from where it was native. In order to operate safely, the massive cargo ships pump in ballast water to maintain constant and equal weight distribution, and also to compensate for fuel and water consumption during the days and weeks it takes to cross an ocean.

Figure 2.17: Zebra Mussel

Unfortunately, when taking in the ballast water, they also take in bacteria, microbes, the eggs, cysts, and larvae of various species, and small invertebrates like the zebra mussel. Changes have finally been made as to when, where and how ships exchange ballast water to reduce the potential for invasive species transfer, but this was occurring for more than a hundred years before the ecological problems caused by the practice were recognized. And even then it took many years to negotiate an agreement on how to prevent further transfer of invasive species. Consequently, many species like the zebra mussels have taken hold in new ecosystems.

In their native ecosystem, the zebra mussels are just one of many species helping to maintain the homeostasis of their home habitat. In the Great Lakes, however, there is nothing evolved to prey upon them. Fish can eat them, but the zebra mussel is low in fat and the shell offers little nutrition, so it requires a great deal of energy to crush and eat the mussels, with little nutritional payoff. Additionally, the zebra mussel is an incredibly efficient filter feeder, and feeds quite well on the native bacteria and microorganisms in the water. This is the perfect setting for an exotic to become invasive: it has ample food resources but almost nothing to limit its population growth. Lake Michigan's crystal clear water, while beneficial to snorkelers, is indicative of an unhealthy ecosystem. The high level of clarity is a result of the zebra mussels consuming most of the microorganisms. The mussels are such efficient consumers in this new habitat that they have outcompeted and replaced the native species that previously fulfilled that role on the food chain. Consequently, the larger native fish have lost their primary food source and have decreased greatly in numbers. The increased water clarity can also increase the sunlight that reaches greater depths and trigger an algal bloom.

Another example of an invasive species is the European buckthorn, *Rhamnus cathartica*, brought in as an ornamental garden tree. Unfortunately, not many North American animals browse on the tree, which otherwise might limit its population. Birds that do eat the berries get diarrhea from the fruit, causing the birds to disperse the seeds over a wide range. Buckthorn has a longer growing season than many native trees in North America. The confluence of these factors: longer growing season, lack of predators, and ease of seed dispersal, has given this species an advantage over the native trees. Buckthorn is not a large tree; it grows under the cover of the larger trees in a forest such as maples and oaks. Once introduced to an ecosystem, buckthorn quickly becomes the dominant species of tree in the understory of a forest. This prevents saplings of the larger trees that define the kind of forest it is from reproducing, quickly changing the forest from an ecosystem in which the maple tree might have been a keystone species to one where the buckthorn is now the keystone species. As we have seen before, the keystone species greatly affects all the other organisms that can live within that ecosystem. As a general rule, at the end of the growing season in late fall, if one type of plant is still green and most others have turned brown or lost their leaves, then there is a good chance the green one is not native to that habitat.

The buckthorn and zebra mussel are just two examples of problem invasive species in the Great Lakes region where I live. Sadly, there are countless other examples that have affected almost all ecosystems around the planet. The impact of invasive species in the middle of a large continent such as North America is much less severe than on isolated islands such as Hawai'i and New Zealand. These small island ecosystems usually lack large predators and have much less "ecological space" to absorb exotic species. Therefore one new species, such as a Norway rat, *Rattus norvegicus*, having no predators, will see an explosion of population as it easily consumes the eggs from flightless birds common to such islands. Whether transported purposefully or accidentally, depending on the size and complexity of the ecosystem and the organisms and niches the exotic species takes over, the introduction of exotic species to an ecosystem always changes the homeostasis of that ecosystem.

The first "P" in the HIPPO acronym represents pollution. Any contaminant put into an ecosystem is pollution. The excess nitrogen runoff from garden fertilizer causing the algal bloom is pollution. In addition to excess fertilizer, there are many other kinds of pollution. When one thinks of pollution some obvious examples come to mind: litter, oil spills, smog, and dirty lakes and rivers. Less obvious are the chemical pollutants that are used so readily to control pests, which are called "biocides." "Bio" means life, and "cide" means to kill. The most common kinds of biocides are weed killers and pesticides used to kill insects, or insecticides. These biocides are often intended to kill one specific pest on a farm, in a lawn, or in a garden, but they also kill many other species.

One such example of chemical pollutants that have had potentially devastating effects on our ecosystem are neonicotinoids. This is a new kind of insecticide. Instead of being sprayed around crops to directly kill the invading insects, it is sprayed on the plants which then absorbs the chemical and distributes it throughout itself making the plant itself toxic to insects. Initially this method of using an insecticide might seem better for the environment as there potentially could be less excess chemical that runs off into groundwater and also does not indiscriminately kill all insects, but instead affecting only those that actually eat the crop the farmer is trying to protect. Unfortunately, these neonicotinoids also make the nectar in the flowers of these plants toxic. When the bees eat the nectar they ingest this toxin. The bee pollinators are not the intended "target" of the toxin. In fact considerable amount of our food is dependent on bees for pollinating the plants to make the fruit. During the past 20 – 30 years there has been great concern over the increasingly prevalent dying off of bees, called "colony collapse disorder." There is increasing evidence that these chemicals can directly harm the nervous system of some bee species and/or possibly make some bees more susceptible to parasites and disease.

In addition to killing other insects or plants indiscriminately, biocides also create another issue called bioaccumulation. The excess insecticide eventually ends up in the water and then plants making up the base of the food chain absorb it. Most living things do not break down these chemicals because they cannot be used for energy or even as nutrients to make new protein molecules or other components of their cells. Instead, the toxins are stored in their original form in the animal's tissues. When the organism is eaten, the chemical is passed on, intact, to the next organism in the food chain. Remember that as we move up the food chain, or trophic-level pyramid, fewer individuals exist at each level. Because the chemical does not break down, it accumulates at higher levels in each *individual* organism, because there are fewer individuals among whom the pollutant can be spread at that trophic level. At the top of the pyramid, there are so few large predators that each one carries large proportions of the toxin in their bodies, making them sick, or even killing them.

A classic example of bioaccumulation in the United States is the use of a pesticide called DDT (dichlorodiphenyltrichloroethane). DDT was used effectively to kill mosquitoes and other disease-transmitting insects—in particular, insects that transmit malaria and typhus. Widespread use of DDT began in World War II to protect troops from the spread of these two diseases. After the war, DDT was widely used in the United States to control the spread of these diseases among the civilian population in the 1950s and 1960s. Unfortunately, this chemical accumulated in aquatic ecosystems, eventually ending up in the top predators. Large raptors, such as bald eagles, ospreys, and condors, were greatly affected by this chemical because DDT caused their eggshells to be so brittle that most of the eggs did not survive. Consequently, these species nearly became extinct in the 1950s and 1960s. Largely due to the research and work of ecologist Rachel Carson, presented in her book called *Silent Spring*, the damage was recognized and DDT is now banned from any use in the United States and all agricultural use worldwide. Because of its effectiveness at limiting disease-carrying insects such as mosquitoes, it is still used in some locations as a means to control the spread of malaria. This is an example of the difficulty that we face with our modern ecological identity. How does a society balance the needs of individuals (those saved from malaria in this case) with the long-term effects of damaging the overall homeostasis of the ecosystem?

PCBs (Polychlorinated biphenyls) and mercury are other examples of chemical pollutants that accumulate in an ecosystem. The United States banned PCB, which is toxic to humans, in 1979. Before then, however, it was a common ingredient in oils and lubricants. Mercury is a waste product from coal-fired power plants. The mercury ends up in aquatic ecosystems and builds up in the fish in that ecosystem. Predators of those smaller fish can accumulate mercury in their flesh. When eaten, mercury damages the kidneys and neurological systems of vertebrates, including humans. Because of this, most locations in North America have "fish consumption advisories," and many types of seafood also come with warnings to consume only certain amounts per week or month.

The second "P" in HIPPO stands for population (or more precisely, overpopulation). Two thousand years ago, the human population numbered around 300 million. Around the year 1800 it crossed the 1 billion mark. While it took about 2,000 years or so to increase the population by 700 million, it then took only 127 years, from the year 1800 to 1927, for the population to double to 2 billion. After only 33 years, the population grew to 3 billion, and then to 4 billion only 14 years later. By 1987, there were 5 billion people on the earth. That number increased by another 2 billion from 1987 to November 2011, when it was estimated that the world population crossed the 7 billion mark. This is an example of exponential growth that will be examined in much more detail in the next chapter. Suffice it to say, as the number of people on the planet increases, the more each of the other factors causing biodiversity loss are magnified.

The last letter in HIPPO stands for overharvesting. Overharvesting is quite simply the exhaustion of natural resources, such as timber from a forest or fish from the ocean. Humans remove about eight percent of the primary production from aquatic systems each year. Some regions of the ocean are much harder hit. Warmer regions have about 25 percent of the fish removed each year, and temperate regions have about 35 percent of the productivity removed each year. Reducing the population of fish within aquatic ecosystems at these levels is greatly reducing the biodiversity of the planet. Overharvesting, especially of oceanic fish, is a problem that is growing exponentially, not only as the human population expands, but also as more and more people eat fish because it is a lean, healthy source of protein. These are but a few examples of the impact our current ecological identity, which includes Habitat destruction, Invasive species, Pollution, Population and Overharvesting, has on our local ecosystems and the biosphere as a whole.

The early beginnings of our kind on this planet were as creatures evolved with adaptations for open spaces, with an ecological identity centered on gathering food, hunting, and possibly even scavenging—though that idea is uncomfortable for many to consider. Certainly we were a potential prey species to large mammal predators on the African savanna. Throughout our evolution, our species' role in our ecosystems certainly changed as technological advances allowed our ancestors to spread into new habitats and ecosystems. Of course it would be tempting to label our species as an exotic species in all of these other new habitats, and even possibly an invasive species. But is our spread, made possible by the adaptation of a large brain, part of our natural evolution?

Now that we have spread throughout the planet, our impact on the ecosystems we inhabit is undeniable. How to respond to these human-caused threats to biodiversity is a difficult question to answer. Many people argue that we should preserve and conserve as much as possible and actively try to fix and manage the ecosystems in which we live. Others argue that the systems are too complex to manage effectively, and our best option is simply to reduce our own impact and let the natural systems repair and maintain themselves. Others still, argue that the world's resources are there for us to use as we see fit, as the dominant, most intelligent species on the planet, and we should not have to sacrifice our own standards of living, jobs, and economy for the other species on the planet. Regardless of the philosophical approach one takes, it is evident that humans play a key role in the ecosystem, greatly affecting biodiversity and changing the niches other organisms fill. In many ways humans have become the keystone species not just for the ecosystems in which we live, but for the *entire* biosphere.

Exploring Biodiversity Web Page
www.exploringbiodiversity.com

Chapter Two Web Resources
http://www.exploringbiodiversity.
com/#!chapter-2-student-resources/cpxo

The key problem facing humanity in the coming century is how to bring a better quality of life—for 8 billion or more people—without wrecking the environment entirely in the attempt.

> -Edward O. Wilson, scientist and Pulitzer
> prize winning author

Chapter 3: Populations

What is your role in a dynamic population?

1 – Counting Populations

Close your eyes and take a moment to return to our imaginary park. Try to visualize the variety of living things interacting around you. There are countless grass plants; maybe a red-tailed hawk, *Buteo jamaicensis*, circles overhead and screeches, searching for its next meal; a red squirrel, *Tamiasciurus hudsonicus*, aware of the circling shadow of the predator, darts to the base of a red oak tree, *Quercus rubra*. American robins, *Turdus migratorius*, hunt for worms while thirteen-lined ground squirrels, *Spermophilus tridecemlineatus*, poke their heads from holes in the ground. How many of each kind of species represents a population? What might be the easiest way to count them, observe how they interact, and determine how and why those populations are dynamic and always changing?

There are two ways of counting populations: census and sampling. A census is a direct count of all individuals in an area. A census is the best way to count a species that is easy to catch or see. Therefore, it is typically used for slow-moving or stationary populations, smaller populations, or a population contained to

93

a small area. A census, then, might be an effective means for counting the number of oak trees in our imagined park. The United States government conducts a direct-count census of the U.S. population every ten years. This is a very labor-intensive method for counting the U.S. population, as people move around a lot more than oak trees! Might there be an easier method to count the U.S. population?

Now imagine trying to count the population of squirrels in the park, or robins, or something as small and abundant as the individual grass plants. This would be virtually impossible. An *indirect* count method would be much more practical. Taking a sample is one way to indirectly count a population. How might you sample the grass population and then mathematically estimate the total population in the park? It would be easy enough to divide the park into smaller segments, called quadrats (Figure 3.1). Using a quadrat creates an area small enough

to manage counting every individual grass plant. Then the researcher can multiply by the number of quadrats to estimate how many grass plants would be in the entire park. Where might errors occur in this method? Imagine you selected an area that was particularly dense, in grass,

Figure 3.1: Quadrats

or instead one that was repeatedly trampled by foot traffic. Using these areas as a representative of the larger population might give you an inaccurate estimate. Therefore, it is important to take several samples to improve the accuracy of the data collected.

How about the squirrels? Marking a grid and counting the squirrels in just that quadrat might be inaccurate with a highly mobile and fast-moving animal. For animals like squirrels, birds, butterflies, fish, etc., scientists use a technique called capture-mark-recapture to estimate population. After capturing, marking, releasing and then recapturing the animals, a mathematical estimating process called the Lincoln Index is used to estimate the population size. The population total is estimated by multiplying the number of individuals captured and marked in the first sample by the number of individuals captured in a second trial, divided by the number of marked organisms captured in the second trial.

$$\text{Population} = \frac{\text{(total caught and marked first time)} \times \text{(total caught second time)}}{\text{(marked animals in second catch)}}$$

For this method to be accurate, it is important to capture and mark at least 20 percent of the population initially, otherwise the chances of recapturing an animal are too small. If you only capture a tiny fraction of animals from that population, you most likely will not recapture any. If you put a zero into the formula above, your estimate of the population would also be zero, which of course is not accurate. Imagine the reverse were to happen and you keep recapturing the same animal over and over again. The formula would calculate to a population of one. This could be an accurate estimate, or it could mean that you are using the mark and recapture method for an animal that doesn't move around much— like a tree sloth. Recapturing the same animal

Video: Lincoln Index https://vimeo. com/167570494

repeatedly would accurately tell you the population in that very small range (perhaps one tree), but not for a wider area, like a whole forest.

Recall the keystone species of kangaroo rat in the Chihuahuan desert. This small mammal is a perfect candidate for indirect population counting using the capture-mark-recapture method. The rats are easily caught in live traps that are about the size of a tennis ball can, but square in shape instead of round. The animal is attracted to bait inside the trap, and when they enter they trip the door, causing it to shut behind them. With the first trapping, scientists insert a microchip under the rat's skin, in the same way some pet owners do with their dogs and cats. Thereafter, when a kangaroo rat is caught, it can be identified with an electronic reader that picks up the signal from the chip. From this data, the researcher identifies where and when the kangaroo rat spends its time in that habitat, and collects enough samples to accurately estimate the population utilizing the Lincoln Index method. Researchers sometimes use less technological marking devices. For example, larger animals are marked with ear tags and neck collars. Smaller animals like birds and rodents are often adorned with leg bands. When not attempting to keep track of specific individuals, but simply to count the number of individuals captured, markings as unsophisticated as fingernail polish on the back of a mouse will work to complete a capture-mark-recapture study.

The U.S. government conducts a direct-count census every ten years. How could the task be accomplished with the less labor-intensive method of indirect sampling? Some statisticians argue that taking a sampling of the U.S. population would be more accurate than physically counting each and every individual. The U.S. Census Bureau does conduct some sampling as a part of their data collection, though not for counting the population. Starting with the 1940 census, approximately five percent of the U.S. population completes a longer census form, which provides the U.S. government with additional information about the population beyond how many people there are and where they live. This allows for the gathering of more detailed information, but does not have the added labor of asking every household in the U.S. to answer a lengthy census form. Whatever method used for completing the census, I think we can agree that capture-mark-recapture would not be acceptable!

2 – Dynamic Populations

In our park, you hear the chattering of squirrels and scratch-ing of their claws. Your attention is drawn to a few trees a few meters off of the walking path where two squirrels chase each other up, down, and around the trunk of a tree. In addition to the number of organisms living in a population, the characteris-tics of each population are also important to understanding the relationships among those populations. Population density is the measure of how many individuals are in a designated area, such as the number of organisms per square meter for small organisms, or the number per square kilometer for larger ones. The denser the population, the more the individuals have to share, or compete for, resources.

Organisms move around to get resources like food and water. They clump together for behavioral reasons like finding mates, caring for young or increasing safety from predators. While density describes the number of individuals within a given space or ecosystem, distribution is how the population gathers within that ecosystem. Large herding prey species like reindeer, caribou, or zebra often distribute in a clumped pattern. Large predator species are more likely to distribute evenly in an eco-system so they do not have to compete with each other for prey, which for a predator is often hard to come by.

What factors might cause the population size and density of an organism in our imagined park to change? Four "rates" di-rectly affect population size. Deathrate measures mortality, or the rate at which organisms die in a given time period. Consider the kangaroo rat in the Chihuahua desert. Dry seeds are its main food source and it is preyed upon by mountain lions. What factors would increase its deathrate? A drought that reduces the grasses' seed production would certainly increase starvation among the kangaroo rat population, and an increase in mountain lion population would certainly increase the deathrate of the rat. A food shortage in a nearby habitat might cause more predators to move into a new area. The addition of a new predator such as a hawk might increase predation on the kangaroo rat, but might also then affect the mountain lion's population density and size. As you can begin to see, the dynamics of a population are more complex than just direct predator-prey relationships between two species.

The second "rate" is birthrate, which is simply the number of individuals born in a given time. The same factors that affected the deathrate of a species might also affect the birthrate, but in the opposite direction. Going back to the kangaroo rat example, an increase in rainfall would increase the grasses' seed production. This increase in food supply might lead to healthier, well-fed mother kangaroo rats that will produce more offspring, thus increasing the birthrate. An increase in grass population and density will provide additional cover from overhead predators, increasing the female kangaroo rat's life span, and allowing more time to mate and have more offspring. Any factor that allows adults to live a longer lifespan or increases the health of the mother during her reproductive years increases the birthrate in a population, thus leading to a population rise. If birthrate and deathrate are

Video: Population Dynamics *https://vimeo. com/167571479*

equal, the population remains unchanged. One can then imagine how population density and competition for resources might greatly influence the birthrate or deathrate.

The third and fourth "rates" are emigration and immigration. Changing conditions might force individuals to move into (immigration) or out of (emigration) a habitat. Emigration is the rate at which individuals move out of an area. Immigration is the opposite—when an individual moves into an area and joins a population. Organisms typically will not emigrate if there is adequate space and resources where they live. Increased density, however, might force some to *emigrate* from their original location and *immigrate* into another location.

This movement of individuals from one habitat to another is the dispersal of a species and describes how a population spreads from one location to other locations. Many of the same factors that affect the four "rates" of population, as well as the density and the distribution also affect the dispersal of a population. Individuals disperse as they search for the easiest ways to get more resources, find shelter, find mates and escape predators, along with a variety of other reasons. These motivations affect the "range" of species. The range is the normal distribution of a species throughout a large area—larger than an ecosystem, maybe even across many biomes.

Your attention is drawn away from the squirrels by a bright, cheery call of a Northern cardinal. To most individuals, singing songbirds are simply background noise, often sounding all the same. But the song of the male cardinal is distinctive. It is a loud, clear song. The male sings the same song over and over again to claim his territory and attract a female. Putting words to the song, one can imagine the male singing, "look heeeere, look heeere, here I am, here I am." The cardinal is a very common songbird in the United States. Due to its large size (for a songbird) and coloring—bright red with a tuft of feathers on the back of its head—most people easily recognize it. The cardinal's success is due to its adaptability, and in particular, its adaptability to human habitats. It is especially skilled at finding food and cracking the types of large seeds common to many backyard bird feeders. Because of this adaptability, coupled with the availability of food provided by humans, cardinals have spread and dispersed northward throughout the United States. Individual birds from more southern populations effectively emigrated out of those habitats and immigrated into regions further north (Figure 3.2).

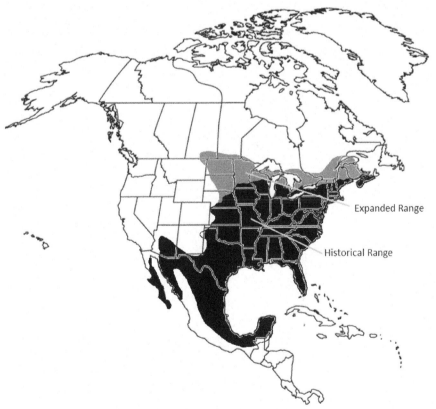

Figure 3.2: Northern Cardinal Range

Two complementary conditions might have facilitated spreading of the species into new ecosystems and even different biomes. Humans provided bird feeders, increasing the birthrate and survival rate of the cardinal population. As the population grew, so did population density, thus increasing competition for resources. When there is competition for resources, individuals have one of three choices: out-compete other individuals, die, or emigrate.

This spread and dispersal of cardinals is not to be confused with migration. Migration is the seasonal movement of individuals or entire populations. Many bird species migrate north in the summer and south in the winter. Monarch butterflies, *Danaus plexippus*, migrate from the temperate regions of North America to Mexico (Figure 3.3). Larger animals, like wildebeests in Africa, migrate as well—though not because of summer and winter changes, but in response to rainy and dry seasonal changes.

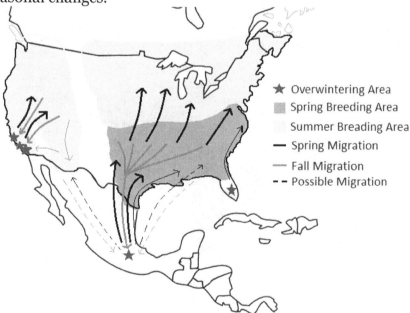

Figure 3.3: Monarch Butterfly Migration

Of course, individuals and populations can only emigrate (or immigrate) if they have the ability to move and if nothing blocks their movement. Geographical barriers, like a river, lake, or mountain, might block an individual's movement from one location to another.

Isle Royale is an island in Lake Superior that hosts a population of wolves and moose (Figure 3.4). The island is too far from the mainland for the animals to swim to it, so how did these large mammals come to populate this island? Depending on the severity of the winter, the lake sometimes freezes between the mainland and the island, allowing animals to move between mainland and island populations. However, they will only do so if there is a reason. The average moose does not simply go exploring! In-

Figure 3.4: Isle Royale

creased pressure from predators during a particular winter or increased competition for food or mates might have motivated some individuals to leave one population and move to the other. Recently, due to increased temperatures reducing the frequency of ice formation between the mainland and the island, the populations of wolves and moose have become isolated. The wolf population on Isle Royale in 2013 was down to a few individuals, not enough to prey upon the moose population. There is debate among the scientific community as to the next course of action—let the situation play out or intervene by adding more predators to the island.

Adaptations and behavior might also influence whether an individual immigrates to a new location. Thirteen species of finches known as "Darwin's Finches" live on the Galapagos Islands. How there came to be thirteen species on the islands, originating from a just one species, will be discussed further in Chapter Six, but for now let's look at how they came to emigrate from Ecuador to these islands. The islands are a few hundred miles from the mainland and are therefore geographically isolated (Figure 3.5). Finches are three to four inches long and not capable of soaring such distances over the ocean like larger sea birds. They do not have the physical or behavioral adaptation for such sustained flight. However, consider an incident where many of these birds were blown out over the ocean in a storm and found refuge on these islands. They did not have the adaptations to find

Figure 3.5: Galapagos Islands

their way back to the mainland, so in a sense, they were marooned on the islands. Because they can fly short distances, they could navigate between the individual islands that make up the archipelago of the Galapagos, but did not have the adaptations to migrate all the way back to Ecuador.

The Galapagos archipelago is a collection of several different small islands, many with slightly different habitats. Some individuals from the original population had adaptations allowing them to survive in some of the island habitats. Those without behavioral or physical adaptations did not survive. If no available niches had existed for any of the individuals to make a living on these islands, they would have all died. A population has to have the necessary adaptations to survive in that new habitat (to spread into it) *and* there has to be an available niche for them to fill—that role and "space" in the ecosystem cannot already be occupied with other species making a living in that exact same niche, unless it has an adaptation by chance that makes it even better suited for that habitat than the native species. In this case it would actually out-compete the native species and possibly take over that niche.

3 – Limiting Factors and Carrying Capacity

The ecosystem in which a population lives has finite resources. Previously I equated the amount of energy coming into the Earth system to a balloon filling up. Only so much can fit. Well, the same is true for the "space" available for a population in an ecosystem. Imagine yourself on that path in the park again. Look around your imaginary park and see the space of the grass, trees, shrubs, and landscaped areas. This habitat is surrounded by an urban landscape (which is a different kind of habitat). Only so many individuals can fit. So what dictates how many individuals of a population can live in an ecosystem? There are many factors that limit a population's size such as deathrate, birthrate, immigration, and emigration rate. These are called limiting factors because they limit the population. The combined effect of the limiting factors is to set the species' carrying capacity within an ecosystem. Carrying capacity is quite simply the number of organisms of a species that can successfully live in an area. The population cannot exceed the carrying capacity of the ecosystem—at least not indefinitely.

Video: Carrying Capacity
https://vimeo.
com/167571479

There are two categories of limiting factors: density-dependent and density-independent. Remember that population density is the amount of population within a given area, often measured as number of individuals per square meter or kilometer. A density-independent factor reduces the population indiscriminately, despite the density of the population. Forest fires, newly-introduced diseases (though communicable diseases may be density-dependent), floods, or other such natural disasters are density-independent factors. The effect on the individuals in the population is generally the same for all of them, no matter if clumped tightly together or widely distributed. In some cases one of these factors usually considered density independent can affect individuals differently depending on the size of the population and if that limiting factor also then greatly increases competition between the individuals for survival. It can then be considered density dependent. Consider a herd of animals rushing to flee a fire. If the population is very densely packed, some individuals may get trampled, pushed aside, or held back from fleeing as they all scramble to escape the flames, compared to a low-density population, where all individuals had plenty of room to run ahead of the flames.

While the density-independent factors tend to be major events or sudden changes, density-dependent are often conditions that are what you might consider "background" conditions also exerting pressure on the survival of the individuals in the population. Density-dependent factors tend to have greater influence on the homeostasis of a population within an ecosystem. The amount of food available is the most important density-dependent factor. If given ample food, a population will always increase. Think back to the previous chapters and the flow of energy. If energy is available, life will transform and utilize it, creating order and more life. More food always equals more population. Scientists studying medium ground finches, *Geospiza fortis* (Figure 3.6), on the Galapagos Islands over the last several decades have witnessed a prime example of this

rule. During the study, there have been
periods of extended drought conditions.
During these times, the plant life on the
island decreases, thus decreasing seed
production. As food dwindles, birds are
in stiffer and stiffer competition for the
few remaining seeds. Consequently,
large portions of the population starve.
This increasing deathrate reduces the
population density to match the new

Figure 3.6: Medium Ground
Finch

levels of available food. Had there been a low population density
prior to the droughts, none starve. At other times the opposite
occurs and it rains for weeks on end. The food source greatly
increases, and the population balloons on the island to match
the available food resources.

Another example of a density-dependent factor is the ability to
avoid a predator. When I was a kid, the bullfrog, *Rana castebi-
ana,* once had a very high population density in the lake where
my family has a cabin in Northern Wisconsin. This time in my
childhood at the family cabin is a key aspect of my own ecological identity. The population was so dense that my older brother
and I could easily wade into the water and scoop up these six to
ten inch frogs with our bare hands or canoe paddles. These are
big frogs easily seen by bald eagles, *Haliaeetus leucocephalus,*
large fish like the northern pike, *Esox lucius,* raccoons, *Procyon lotor,* and other predators—or young boys! After dark, the
male frogs call to attract mates. A low population density would
be an advantage when competing for food or finding a mate,
but a disadvantage for avoiding predators, which might seem
counter-intuitive at first. But consider this: while being widely

spread out would allow mates to locate one another when they make their long drawn-out "rummm" croaking noise, it would also make it easier for a predator to find them by honing in on individual's singing. However, with a high population density, they all call at the same time, which confuses the predators, or at least decreases an individual's chance of being the one that is caught. There is safety in numbers. Each frog's call is slightly different, so even though they all sound the same to larger predators (or us), the individual bullfrogs can distinguish one from another.

It is common to think of predators as "bad" for the prey population. After all they do eat them! However, the homeostasis of both the predator and prey population is dependent on the predator-prey relationship. As the predator species' population rises, the prey species' population will drop. When prey numbers decline, there is less food for the predator, and the predator numbers also decline. More starve as competition for food intensifies. As the predator population drops, the prey population will recover. The cycle continues, often reaching a balance between the predator and prey species.

To illustrate the predator-prey relationship, we can return to our Isle Royale example of the wolves and moose. Scientists have been studying the relationship between these two animals for 50 years. The Isle Royale wolf and moose populations are linked together, which you see by studying the pattern in the graph (Figure 3.7). There are other factors that also affect the populations. At times the wolf population has suffered from density-independent factors such as a disease that affects the small population of wolves on the island. This happened around 1980. This sharp, sudden drop in the wolf population

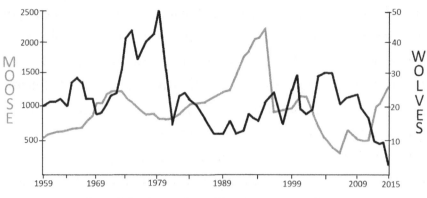

Figure 3.7: Isle Royale Wolf and Moose Population

was followed by an historic increase in the moose population. Ultimately, the lack of predators was not good for the moose either, as the increased moose population took a heavy toll on the island vegetation. This is another example of a density-dependent limiting factor. In 1996, nearly 80 percent of the moose population died from lack of food as the increased population decimated the available food sources. A healthy predator population keeps the prey population stable by keeping the numbers in check. In this case, the reduction of the wolf population caused the moose population to increase beyond the carrying capacity of the Isle Royale ecosystem. The wolf and moose population has never been stable since the crash of the wolf population in the late 1970s and it is likely that soon there will be no wolves left on the island. This lack of predation will result in continued cycles of exponential moose population growth followed by population crashes as they exhaust and trample the vegetation faster than it can regrow to support the elevated population.

4 – Human Population

Here is a scary question you may not have previously considered: Do these factors that we can clearly see affecting populations of other species also apply to the population dynamics of the human species, or have humans successfully removed themselves from these natural laws? Could our species exceed the carrying capacity of planet Earth and suffer the same fate as the moose on Isle Royale?

The human population, *Homo sapiens,* began in Africa around 200,000 years ago. The population of our species was once possibly as low as a few thousand. We were nearly extinct! During the past 200,000 years, our species has proven quite adaptable and now humans can live in any habitat and ecosystem—even outer space. Adaptability enabled by our intellect seems to be our greatest adaptation for survival. This is an important part of the story of our ecological identity. Our species and our ancestors may have evolved in Africa, but we didn't stay there of course. The first of our ancestors to leave Africa was most likely a species called *Homo erectus* or *Homo ergaster* about one million years ago. Spread throughout Europe and Asia were others from our family tree, including *Homo habilis* and *Homo neandertalensis. Homo sapiens* emigrated out of Africa 100,000 to 125,000 years ago—for a combination of all the reasons described earlier—and by anywhere from 15,000 to 30,000 years ago, individuals from our species had spread throughout all of Africa, Europe, Asia, Australia and the Americas.

At the end of the last ice age our species' ecological identity underwent a major shift. Twelve to sixteen thousand years ago, when the ice age ended, grains such as wheat, rye, and barley became much more available as a food source. Not only was this an easy-to-gather food source, it was also easy to store for long periods of time without spoiling. People began living together in villages as grain gatherers. When the grains in an area were depleted, the village would move. Prior to this, humans were primarily hunter-gatherers and lived a nomadic lifestyle dependent on plants they collected and animals they hunted. The grain gatherers were still nomadic at times, but now were beginning to settle in more permanent villages—at least until the surrounding plants and animals were depleted.

Humans living in a region of the Middle East referred to as the Fertile Crescent (Figure 3.8) developed the beginnings of western agriculture approximately 10,000 years ago, and thus made the final transition from moving-the-village-to-the-food, to bringing-the-food-to-the-village. The ability to stockpile food allowed for

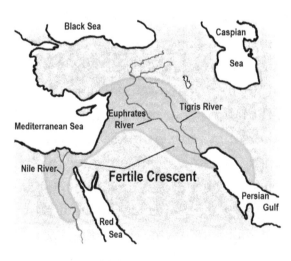

Figure 3.8: Fertile Crescent

the beginning of a population increase beyond the carrying capacity of local ecosystems. The increased population density caused the humans living in those areas to begin spreading and dispersing, just as we saw earlier with the cardinal. More food equals more population, which means greater competition for resources, thus increasing the need for the population to emigrate and disperse to new regions, thus increasing the range of the species.

The population of humans living in the Fertile Crescent was not the only group of humans on the planet, as the human species had already spread across much of the globe during warm periods between ice ages; but even still the global human population was relatively low at this time in our history. Many of the populations of humans living outside of the Fertile Crescent had also developed some agricultural methods, but these tribes relied primarily on hunting and gathering for acquiring food. The humans that developed agriculture in the Fertile Crescent dispersed and immigrated into new locations. They spread east and west, staying within the same climate, and therefore could move their crops and animals with them. The best climate for agriculture is found in the temperate grasslands and forests. These seasonal climates tend to have better soil and habitats for many of the crops that humans came to be dependent upon, as well as better habitats for the animals used for work (beasts of burden), like oxen and horses. As a result, these budding agriculturalists spread throughout the globe into temperate regions, bringing the agricultural way of life with them—usually at the expense of the tribal peoples that already lived in those regions.

Video: Human Population and the Agricultural Revolution *https://vimeo. com/167572662*

This dispersal resulted in incremental human population growth during the 10,000 years after the agricultural revolution. Within the past 200 years, however, the human population has grown exponentially, reaching 7 billion people in 2011 (Figure 3.9). The majority of people live in Earth's largest cities, which tend to be on the coasts. This is particularly true in North America and Eurasia, where the population density near the oceans is quite high and in the interior of the continents it is much lower. What ecological impacts might this disparity of population density have on these areas' natural resources? As population growth continues, will these different regions face different ecological issues due to increased population density?

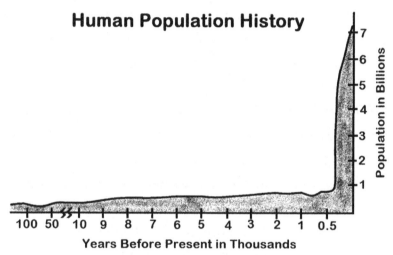

Figure 3.9: Human Population History

Fortunately, the rate of global population growth has begun to slow because fewer world regions are impoverished. It might seem a bit counterintuitive, but increasing wealth generally decreases birthrate in a population. As average income and access to resources and education increase, individuals tend to have *fewer* children and also tend to have children later in life. With organisms such as small mammals, increasing the health of the mother increases litter size and birthrate. Though humans and other larger mammals do not have a "litter," the health of the mother does affect the ability to get pregnant, carry a child to full term, and give birth to a healthy offspring. Humans, though still a part of the eco-system, have changed the rules somewhat for themselves. Impoverished individuals tend to have *more* children and begin to have children at an earlier age, even though more of their children die before reaching adulthood. The birthrate is therefore higher in developing and impoverished regions of the world.

Increasing the literacy of girls allows them to stay in the education system longer, delaying when they begin a family. Postponing childbirth greatly reduces population growth rates. Because humans have such a long gestation period and child-rearing takes 15 to 20 years, delaying when a mother begins having children greatly reduces how many children she will have. With increased wealth, women also decrease the number of children they have. If the children are healthier and more survive into adulthood, parents also have fewer children and therefore the total birthrate is low-er than in impoverished countries where the women do not have access to education and monetary resources.

Though the global birthrate has declined greatly in the last 20 to 30 years, the global population is still rising due to something called population momentum and is estimat-ed to reach 9 billion in the next few decades. There are such a large number of individuals capable of having children at this current time that even if the birthrate decreases, we will still add millions and millions of people to the population. The trajectory of the population growth is analogous to the trajectory of a rocket shot towards outer space. When the rocket engines are burning, the rocket is constantly accelerating as it tries to escape Earth's gravity. When the rocket engines stop firing, the rocket does not immediately begin to fall back to Earth, but continues to climb due to momentum, until Earth's gravity begins to pull the rocket back down. Eventually, if the birthrate continues to decline, population growth will slow, and the world human population will decrease.

What is the carrying capacity of Earth for the human species? Does the ecological concept that more food equals more population also apply to humans? How has the ability to store, stockpile and transport food around the world changed the human population? Certainly, there must be a limit to how much food can be grown, how much energy is available, and how much space is available. These may be new ideas to you, but they are certainly not new issues. In fact, Thomas Malthus first proposed these concepts of the impact of overpopulation over 150 years ago, and greatly influenced Charles Darwin and his research. Malthus warned that a population unchecked, provided with abundant resources, will always grow to levels beyond subsistence, and that conversely, food resources will always limit population size. We have seen this to be true in populations from bacteria to the moose on Isle Royale. Can and will the same thing happen to the human population?

As we saw before, as human population grows, biodiversity is reduced by the five factors represented with the acronym HIPPO. At what point will we not be able to produce enough food, or at what point will the spreading human population cause such a loss of biodiversity that the ecosystems will no longer maintain homeostasis? These are scary but necessary questions to ask and answer if we are to live sustainably as ecological members of the biosphere.

Exploring Biodiversity Web Page
www.exploringbiodiversity.com

Chapter Three Web Resources
http://www.exploringbiodi-versity.com/#!chapter-3-stu-dent-resources/cktd

Man does not weave this web of life. He is merely a strand of it. Whatever he does to the web, he does to himself.

-Chief Seattle, Ancestral leader of
Suquamish Tribe

Chapter 4: The Balanced Body

How does your body maintain homeostasis within an ecosystem?

1 – Foreign Invaders

I have had you imagine yourself walking a city street, on a playground, and eventually observing all the life around us in a city park. Through that imagery you have explored the homeostasis of an ecosystem and been introduced to the idea that the ecological identity of our species is linked to our role in participating in the homeostasis of the biosphere. Now let's dig deeper into the ways in which your body maintains its own homeostasis and also directly participates in the homeostasis of the ecosystem in which you live.

What if you considered your body an ecosystem unto itself and applied the same kind of systemic thinking that we apply to ecosystems and all of its interconnected workings to the human body? In fact, recent genetic decoding of bacteria residing on and in humans has revealed that each of our individual bodies is indeed its own ecosystem. A very small ecosystem is sometimes referred to as a microbiome. Each one of us serves as a wholly unique microbiome to other organisms. Immediately after birth, bacteria begin to grow on and in our bodies. However, this is not an infectious disease. As we grow, a variety of bacteria grow along with us, and our immune system learns to recognize the menagerie of bacteria living on and in us as a part of our microbiome. Healthy people share their bodies with up to 10,000 different species of bacteria! Many of these bacteria help maintain the health of their human host, perhaps helping to keep in check the smaller number of harmful bacteria living within this microbiome. Maybe this is similar to the predator-prey relationships from Chapter Three. Interestingly, there is not one set of core bacteria on all of us; because of our early exposure to bacteria, we each host a unique microbiome.

Like an ecosystem, the human body also maintains homeostasis. Failing to do so causes illness or even death. Returning to the park, recall the children all touching the same playground equipment and potentially spreading germs. Now imagine this scenario you've probably experienced: you wake up one morning feeling an ache in the back of your neck; your head is pounding and your nose is stuffed. On top of that, you shiver from cold that seems to emanate from inside your body, even though you are tightly tucked under your blankets. Chances are that a germ, such as the flu virus, has invaded you (Figure 4.1).

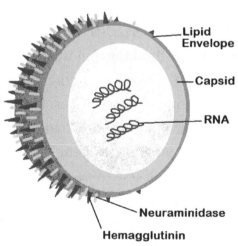

Figure 4.1: Influenza Virus

Unlike bacteria, a virus technically is not a living thing because it does not have cells that burn energy and it cannot reproduce on its own. Instead, it consists of a simple protein structure surrounding a simple DNA or RNA strand. Like a syringe, the virus injects its genetic material into your cells. This is a unique and ingenious reproductive strategy on the part of the virus. Your cells are like little factories that are designed to convert a genetic code into protein molecules. When the virus injects DNA or RNA into the host cell, that genetic material combines with the host cell DNA and is then mistakenly read by the cell. When the host cell does what cells do—read DNA code to make specific proteins—instead of making the protein it was planning on assembling, the host cell makes the protein molecules that make up the structure of the virus. Your cell unwittingly replicates the virus that infected it, and continues to do so until the cell is literally bursting with replicated viruses. As this happens, your cells explode in a process called lysis and new viruses flood your body. These viruses start infecting new cells, which repeat the virus life cycle. Eventually many of the viruses escape from your body (during a sneeze, for example) and infect other people.

Fortunately, in the case of many kinds of influenza, your body usually learns how to destroy the viruses and you recover within five to ten days. The immune system in your blood stream accomplishes this task. In doing so, your immune system learns to recognize the outer protein structure of the virus and communicate to the rest of the body how to make antibodies. Antibodies are proteins that attach to the outside of the "antigen," which is a term for any foreign invader into the body. The antibodies work in a couple of different ways. One way is to bind to and coat the surface of the virus, preventing it from physically entering the cells. Another way is to bind to the outside of the virus as a signal to other proteins, called complements, to break apart the virus, or to cause the white blood cells (leukocytes) to recognize and engulf the foreign invader. During such an infection, it is safe to say that your body is out of balance and no longer in homeostasis. Through the functioning of cells, the interaction among cells, and the interaction between your cells and the outside environment, your body regulates the amount of chemicals, nutrients, water, and all other factors keeping you alive and in homeostasis.

How humans have interacted with the environment during the past few hundred years is part of the story of our immune systems. There is growing research to suggest that we actually have made

Video: Viruses and the immune system *https://vimeo. com/167633320*

our world too sanitary for ourselves. This is called the hygiene hypothesis. The idea is that if our body's immune system normally learns how to fight off pathogens by early exposure, but we eliminate exposure to those things early in life, then we will see an increase in diseases in which the immune system does not react properly to invaders of the body. Studies of siblings show that children who spend more time on a farm, around farm animals, or exposed to many other children in daycare, have lower rates of allergies as adults. It very well could be that our ecological identity involves much more intricate interactions on a cellular level than we ever previously imagined. Maybe a basic tenet of our ecological identity as a species is that in one sense we all really are a world unto ourselves, while also undeniably connected to all the other species.

2 – Movement of Molecules In and Out of the Cell

The homeostasis of your body is dependent on the homeostasis of each of your cells. For the individual cells and the body as a whole to function properly, each of your cells must have the right balance of molecules moving in and out. The cell membrane controls which molecules come and go, including, oxygen, water, proteins and sugars, just to name a few. Ironically, your cells cannot directly control the movement of the most vital molecules (oxygen and water) because they are so small that they can simply slip between the molecules that make up the cell membrane.

To understand this, we must first understand the structure and function of the cell membrane. It is tempting to think of the cell membrane as being like a balloon or plastic bag but that is not an accurate model for the cell membrane. Two scientists, Danielli and Davson, proposed the fluid-mosaic model in 1935 to describe the cell membrane. They were building off the work of other scientists, who had determined with a variety of experiments and observations that nonpolar molecules, or molecules without an electrostatic charge, like O_2, can freely pass through the cell membrane and that therefore it must be porous in structure. Danielli and Davson adapted previous ideas and proposed that the cell membrane was made up of two layers of lipids, which are fat molecules, covered on the outside by proteins, sort of like a sandwich. This was before microscope technology allowed them to actually see the membrane, so they deduced their model purely by observing its chemical content and function.

Twenty-two years later, J.D. Robertson proposed a modified version, this time with the advantage of an electron microscope, which allowed him to get closer to the current model by seeing two distinct layers of the cell membrane. Because the technology was still limited, however, he could not see it well enough, and he concluded that the two layers were covered in a layer of proteins as well, so was still describing it as a lipid sandwich covered with protein "bread." S.J. Singer and Garth L. Nicolson, provided the current model of the cell membrane in 1972. This model, which is really the cumulative work of many scientists over the course of a century, explains the current knowledge of the structure and function of the cell membrane and is the basis for our understanding of how membranes function. Amazingly, scientists understood the structure of DNA, which is inside the cell, before they were able to describe the structure of the membrane surrounding the cell.

The fluid mosaic model is aptly named in describing the cell membrane. Fluid means to be constantly moving, and mosaic means made of many parts. The cell membrane, as stated before, consists of protein and cholesterol molecules embedded in two layers of fat molecules, called lipids. The lipids, which make up the majority of the membrane structure, look somewhat like jellyfish, with a body or head end and tentacle-like tails hanging below (Figure 4.2). How these lipid molecules react to water allows your cell membrane to keep its structure.

The tails of lipids are hydrophobic. This means that they repel water. I remember this by breaking down the word and thinking, "hydro" means water and "phobic" means to be afraid of something. Because water is polar and has a slightly electrostatic charge, the lipid tails and water molecules repel one another.

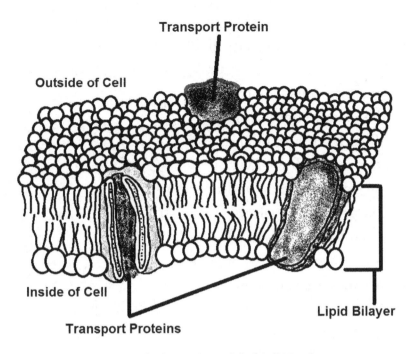

Transport Protein

Outside of Cell

Inside of Cell

Transport Proteins

Lipid Bilayer

Figure 4.2: Fluid Mosaic Model of Cell Membrane

The heads of the lipids are hydrophilic, which is the opposite of hydrophobic, so meaning they are attracted to water. Because your cells have both water inside them and surrounding them, the lipids have no choice but to arrange themselves with the heads touching the water molecules (inside *and* outside the cell) and with the tails repelled, so that the lipids naturally (without any energy by the cell) form two layers with the heads on the outside and the tails on the inside. If you drop a mass of

Video: Cell Membrane
https://vimeo. com/167651210

these lipids in water, they naturally form spherical shapes with the head ends touching water protecting the tails in the middle of the layer. These naturally-occurring structures called "micelles" may have been precursors to the first cells (Figure 4.3).

These individually moving parts that make up the cell membrane create gaps between the lipids. The smallest of molecules, like O_2, CO_2, and H_2O, pass through these small gaps, without the cell having any control over their movement from one side to the other, by the process of diffusion. Even through water is polar and has a charge which repels the tails of the lipids, individual H_2O molecules can slip through while larger amounts of water molecules bonded together repel the tails.

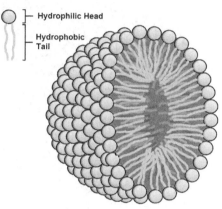

Figure 4.3: Micelle

Diffusion is the process of molecules flowing and spreading out from where there are many of them to where there are fewer of them. No planning or preparation is involved—just simple physics. I find it amazing that my cells rely simply on the laws of physics for the movement of the three crucial molecules essential to my survival. Our cells show an exquisite amount of faith in science!

3 – Movement of Molecules

Up ahead in the imagined park is a popcorn wagon. You see the wagon before you smell the popcorn, but as you get closer to the wagon, the smell eventually reaches you. Molecules from the butter flavoring evaporate into the air and spread from where they are most concentrated (the popcorn wagon) into the open air of the park. The butter-flavoring molecules continue to spread as the moving molecules of N_2, O_2, and CO_2 in the air bump them around. The flavor molecules keep spreading until they are evenly distributed, or equally *diffused,* throughout the park.

The same principle applies to molecules inside and around your body's cells. All of the molecules are always in motion. They bump into each other until spread equally. All molecules spread from areas of high concentration to areas of low concentration. This is diffusion. Think of a bucket of water perched on a partially opened door, waiting for the victim of a practical joke to open the door. The molecules are highly concentrated in the bucket. When they spill on the victim, the molecules move "down" from high to low concentration via diffusion, until equally spread on the floor. We call this difference between the high and low regions the "concentration gradient."

Now let's apply this to the cell. Recall that when completing cell respiration, the cell takes in O_2 and releases CO_2 during this process. As it uses O_2, the amount of O_2 in the cell becomes lower in concentration than the amount of O_2 outside the cell. The concentration gradient of O_2 is higher outside the cell than inside. O_2 molecules will then diffuse across the membrane and into the cell, until it is equal on both sides.

Conversely, CO_2 waste from cellular respiration accumulates in the cell, and the CO_2 diffuses out of the cell into the blood passing by in the capillaries, which are the smallest blood vessels in your body. Each cell is in direct contact with at least one capillary. As blood with low CO_2 concentration passes by the cell, excess CO_2 diffuses from the cell, through the capillary wall and into the blood, which then carries it away. The cell takes advantage of the physics involved in the constant movement of molecules down the concentration gradient to transport and regulate the levels of CO_2 and O_2 within itself, while not having to expend any energy to do so. Very efficient!

The cell uses the exact same process to regulate the amount of water moving in and out as it does with CO_2 and O_2. However, when diffusion occurs with water molecules it is called osmosis. The reason for the difference in terminology is that diffusion describes the movement of molecules that are solutes, like O_2 and CO_2. They are called solutes because they are dissolved in a solution. That solution is water, which is the solvent. So diffusion is the term used for the movement of solutes from high to low concentration and osmosis is the term used for the movement of the solvent, H_2O, from high to low concentration. The two terms are linked because the solutes must be dissolved in a solvent, and the concentration of the water is defined by how much solute is dissolved in it. Water that does not have any solutes in it is referred to as distilled or deionized water. It is 100 percent water and 0 percent solute (like oxygen, carbon dioxide or salt) and is as concentrated as it can be. The only way to lower the concentration of distilled water is to add a solute, salt for example, so that the solution is 90 percent water molecules and 10 percent salt molecules for example. By doing this we have decreased the concentration of water in the solution but increased the concentration of salt in the solution. In the end, diffusion and osmosis happen between cells in your body, and between cells and the blood in your body, without requiring any energy and are, therefore, examples of passive transport.

However, the cell is not completely at the mercy of molecular movement that simply follows the laws of concentration gradients and physics. There are ways in which the cell can control the movement of molecules in and out of itself and between itself and its surrounding environment. Facilitated diffusion is another form of passive transport. As implied by the name, the cell *facilitates* the movement across the membrane, but still relies on diffusion and natural concentration gradients to move the molecules. Facilitated diffusion is how glucose molecules move in and out of some of your cells.

To visualize this process, let's compare your cell to a house. Even a very well-constructed house is not completely air-tight. There are cracks and tiny gaps between components of the house allowing air to leak in and out. Think of the movement of smaller molecules in and out of cells as similar to air leaking through cracks and seams around the windows and doors of your house. The size of these cracks and seams limits the amount of air and prevents larger objects from passing though. If you want more airflow, or you want your dog to be able to get out of the house, you have to open a window or a door. The cell also opens "windows" and "doors," allowing larger molecules like glucose to diffuse through the membrane. While still a form of passive transport because the glucose molecules move down the concentration gradient without requiring energy, the cell does have to open a passageway, called a transport channel, for this facilitated diffusion to occur. It was the proteins making up these transport channels that were initially discovered during chemical analysis of the cell membrane and mistakenly thought to cover the lipids as the "bread" in early models of the cell membrane. In actuality the protein channels are embedded in and span the distance across the membrane, sort of like a tunnel through a mountain, and each protein channel is specially shaped so only the desired molecule, like glucose, can slip through. In our house analogy, the protein channels are like opened windows or doors. When you open a window on a cool day, the cool air flows into the house, but the house did not push out or suck in the air—it just flows in and out depending on the temperature and pressure difference (and the wind) between the air inside and outside the house.

How does a cell move molecules in or out that are too large or too electrostatically charged to fit through the spaces between the lipids, or even through one of the protein channels? What if a cell needs to move molecules *up* the concentration gradient instead of down? Most common with salts, like sodium chloride or potassium, movement against the concentration gradient requires a process called active transport. Active transport is like putting a fan in an open window of your house, forcing air the opposite way it would naturally flow with the breeze. Moving molecules down the concentration gradient is analogous to water flowing downstream or downhill, while moving molecules against or up the concentration gradient can be thought of as pumping water upstream or from a well in the ground.

After reaching the popcorn wagon in our imaginary park, you sit on a nearby bench and eat your bag of popcorn. You can taste the butter and the salt. Salts are necessary nutrients that cells routinely have to move against the concentration gradient, or "uphill," which they accomplish through specially-shaped protein pumps that operate like a lock and key. These pumps are made of interchangeable protein parts called amino acids. The cell creates a protein specific to every molecule needing transport. It is common to think of all the happenings inside of our body as chemical reactions, as things dissolving, burning, mixing, and so on. But in reality, how a protein works is determined by the physical shape of the molecule. Changing the shape of the protein changes the function of the protein—what and how it bonds to other molecules. The cell can quickly build the proper protein pump for each specific task and then take it apart and reuse the protein molecule parts to make another, slightly differently shaped protein for the next task, much like a child can build many different objects with the same set of blocks.

One example of how the cell uses pumps made of differently-shaped molecules is the sodium-potassium pump, which the cell uses to move much-needed sodium (Na^+) and potassium (K^+) ions. By changing the shape of the protein, the cell can actively cause the movement of the molecules to another location—even against where it would normally diffuse—sort of like pushing a ball up a hill or pumping water from a well. Most commonly, cells need to move sodium across the cell membrane or the membrane of organelles inside of the cell. To do this, the cell builds a sodium-potassium pump and embeds it into the membrane, similar to the protein channels used for facilitated diffusion, except the pump is not like a simple open window or tunnel, but more like a hotel elevator that requires the user to swipe a room keycard to get it to work.

The pump is designed to move sodium from one side of the membrane to the other. The sodium fits into receptors on the pump on one side of the membrane, like the keycard activating the elevator. Now the sodium is bonded to the protein. Next the cell uses ATP to physically change the shape of the protein molecule. When an ATP molecule donates a phosphate group to the protein, this changes the protein's shape. When the shape of the protein bonded to the sodium ion (Na^+) is changed, it causes the protein to transport the Na^+ through the membrane to the other side, just as the elevator rider with the keycard is carried up to the suite at the top of the hotel. It almost sounds magical, but really all that is happening is the ATP changes the protein molecule's shape and consequently how it fits into the membrane. This causes the protein to spin around, sort of like a revolving door, in the cell membrane. As it spins it moves the Na^+ to the other side. Now that the protein has changed shape with the addition of the phosphate group, the Na^+ no longer fits into the protein molecule and is released (but now on the other side of the membrane because the protein has spun around). The protein is now shaped perfectly to bind with a potassium ion (K^+). When the K^+ binds with the receptor on the protein, this changes the shape again, causing it to flip back to its original orientation, taking the K^+ with it across the membrane and releasing it on the other side. The protein is now back to its original shape and ready to repeat the process again, like a revolving door (Figure 4.4). The sodium-potassium pump is called active transport because it requires energy to make the ATP in the first place during cell respiration.

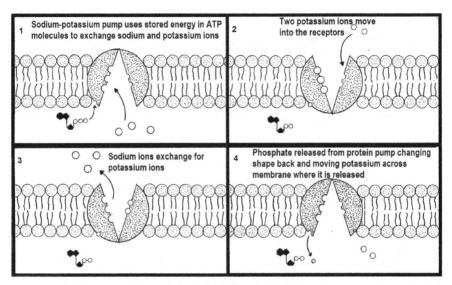

Figure 4.4: Sodium-Potassium Pump

When might a cell need to transport salt? Imagine sitting on the park bench and feeling that the sun is now hot. You wipe away a bead of sweat, but some gets in your eye. It stings because a primary ingredient in your sweat is salt. Maintaining body temperature when overheated requires your body to perspire, placing water on the surface of your skin, so that when it evaporates, it pulls some heat with it and cools your body. Recall that the cells cannot actively pump water, but they can pump salt. The interior of the membrane and the salts have similar electrostatic charges, preventing the salts from diffusing through the membrane like water, oxygen and carbon dioxide do, so the interior of the membrane and the salts repel each other like two positively-charged magnets. Therefore, the cell has to use protein pumps to move salt against the concentration gradient and across the membrane.

Now we come to the connection between the movement of salts and the movement of water. As a general rule, wherever the body pumps salt molecules, water molecules soon follow, but not because the water is attracted to the salt. The connection has to do with why the movement of water molecules is called osmosis instead of diffusion: the water is the solvent in which the salt is dissolved as the solute. When your cells move salt across the membrane and out of the cell, this reduces the solute (salt) in the solvent (water) inside the cell thus increasing the concentration of water inside the cell as compared to salt. The total amount of water did not increase, but the ratio of solvent

Video: Trans-
port of Mole-
cules in Cells
https://vimeo.
com/167903981

(water) surrounding the solute (salt) increased, effectively increasing the concentration of the water. As the solute (salt) is moved across the membrane and out of the cell and added to the solvent (water), this increases the ratio of solute to solvent outside the cell, effectively decreasing water concentration or percentage outside the cell. Water diffuses from higher concentration to lower. Therefore, the cells in your skin actively pump the salts out of its cells, causing the water in the cells to follow the salt and diffuse out and onto your skin, which is why your sweat is salty.

Not only does the connection between salt and water come into play for how your body makes sweat, but these principles affect the hydration levels of all of the cells in your body. If you put a cell with 90 percent water and 10 percent salt into a solution of 100 percent water, the water outside the cell is a higher concentration of water molecules than the water inside the cell, and the system is referred to as hypotonic. The water molecules will diffuse across the membrane until they are equal. If the salt cannot move out of the cell, the concentration can never equalize because there will never be any salt outside the cell. The water will flow in until the cell cannot take any more and the cell literally explodes. This is called lysis, like when a cell reproduces so many viruses it explodes. If you get an IV of fluids at a hospital for dehydration, it is not pure water because the pure water would flood your blood cells and cause them to explode. Instead, doctors use a saline solution, comparable in concentration of salt to water percentages as the fluid in your cells. If the concentrations are the same, then the system is referred to as isotonic. Water will flow into the cells and equalize with the concentration outside the cells. A cell surrounded by an environment where the water concentration is lower than inside the cell will lose water to the environment. This situation is referred to as hypertonic (Figure 4.5).

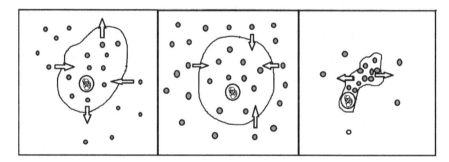

Figure 4.5: Isotonic, Hypotonic, and Hypertonic

Now let's apply these principles to a plant. Near where you imagine yourself sitting on the park bench, there are some planted flowers that look a bit sad and droopy. Despite the recent rain, these plants have begun to wilt. If park maintenance were to set up a sprinkler for these plants, within minutes of being watered, however, they will stand up tall again. Remember that water is one of the inputs in the photosynthesis equation. If the plant does not absorb water through its roots when doing photosynthesis, it will use up its water reserves. As the plant releases oxygen, the cells shrink as they lose water molecules in the production of glucose through photosynthesis. When this happens in an animal cell, the animal cell shrinks down to almost nothing. However, plant cells are surrounded with a rigid cell wall, which give the plant its rigid structure and keep it from shrinking too much. As the cells shrink from loss of water molecules, the plant becomes comparable to a brick wall that is not properly fitted together—loose and shaky (Figure 4.6). However, add water to the plant low in water concentration and the water molecules diffuse quickly into the cell, down the concentration gradient, filling up the cells because the hypotonic solution floods the cells. I remember that this situation is hypotonic because hypodermic needle starts with the same prefix, "hypo," and a hypodermic needle is what a nurse uses to inject medicine into your body. This flood of water into the cells causes them to swell and press tightly against the other cells around it, now causing them to fit together like a well-constructed brick wall.

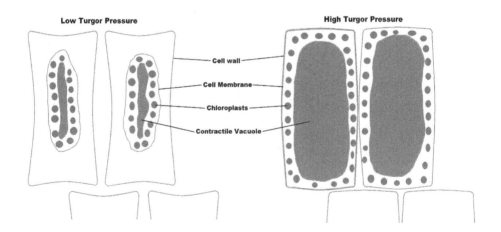

Figure 4.6: Turgor Pressure

The amount of pressure the cells exert on one another in a plant is called turgor pressure. The greater the turgor pressure, the more rigid and stiff the plant becomes. Thinking about how the water moves in and out of the cell based on concentration gradients, now imagine you want cut up celery sticks to remain as firm as possible. Would you leave the celery sitting on the counter exposed to the air, put the celery in salt water, or in pure water? Leaving the celery exposed, water will evaporate out of the cells, causing a loss of turgor pressure and limp vegetables. Putting the celery in salt water will cause the water to flow out of the cell because of the high salt and low water concentration outside of the cell. Put them in water, and water will diffuse into the cells and increase the turgor pressure, making them stiff. Recall this is all passive transport and therefore even though the celery plant is no longer alive, water still moves in and out of the cells. This demonstrates that the cells do not directly control water movement, and that their ability to maintain homeostasis is directly linked not only to their cell functions, but also to the environment in which they live. How might these cell processes connect you to this park ecosystem in which you are imagining yourself, and how might that be a component of your ecological identity?

Video: Turgor Pressure
https://vimeo. com/167651249

4 – *Cells Come From Cells*

The movement of atoms, ions and molecules in and out of a cell is crucial to the cell's participation in maintaining homeostasis of the body. What if your body needs to make more cells? Where do new cells come from? Quite simply, they come from other cells. Imagine eating your popcorn. You take a bite, and then, "Ouch!"—you bite your lip. With your tongue you can feel the cut. In a few days, the cut will be healed and gone. New cells will have grown to seal up the wound.

All cells have a life cycle and a set life span, referred to as the cell cycle. This is the cycle of the cell forming, growing, and then eventually splitting to form two new cells. Scientists have divided the cell cycle into "phases" (Figure 4.7). Interphase is the part of the cycle in which your cells spend most of their time. The greatest portion of interphase is the G_1 phase, or growth phase. The cell grows and carries out its assigned job depending on the kind of cell it is. To ready itself for division, the cell en-

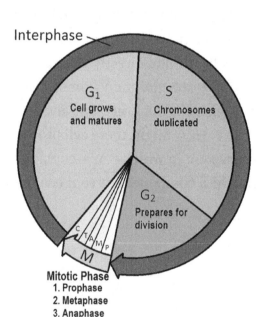

Mitotic Phase
1. Prophase
2. Metaphase
3. Anaphase
4. Telophase
5. Cytokinesis results in 2 new daughter cells

Figure 4.7: Cell Cycle

ters the synthesis phase, or S phase, in which it duplicates all the chromosomes, where the genetic code is stored. After duplicating the genetic material, the cell enters a second growth phase, the G_2 phase. During G_2 the cell finishes making essential proteins and completes final preparations for cell division. The cell proceeds through mitosis, or the M phase. At the completion of this process the original cell no longer exists; instead, two new cells made up from the DNA and parts of the original cell take the place of the one cell.

This idea that cells come from other cells is one of three main tenets of a theory called the cell theory. The second tenet is that all living things are made of cells. The third tenet is that cells make up the basic unit of structure and organization of all living things. The development of the cell theory is an interesting example of how scientific ideas require the participation of multiple voices, and often take many years, to develop.

In the 1600s, Zacharias Jensen was a glasses maker who most likely developed the first microscope. A scientist named Anton van Leeuwenhoek heard about this device and made one of his own. Looking at scrapings from his teeth, he discovered "animalcules." Like molecules are the microscopic chemical components of matter, the word animalcules was used to describe microscopic animals. That is what he named them anyway. What he was most likely looking at were bacteria. Robert Hooke learned about this research and technology and began looking at everything he could find that fit under a microscope. He noticed that cork from oak trees looked like a bunch of little boxes or the small rooms that monks lived in, and coined the term "cells." Fast-forward 200 years, when two German scientists were simultaneously studying living things under microscopes. Mathias Schleiden studied plants and concluded that all plants were made of cells. At the same time, Theodore Schwann was studying animals and made the same observation. This led to the portion of the cell theory stating that all living things are made of cells. Lastly, during the same time period, another researcher, Rudolph Virchow, completed experiments to show that cells did not spontaneously form, but only formed if other cells were already present, thus leading to the conclusion that cells come from other cells.

This process of cells coming from other cells is a highly ordered process; the purpose is to take the one original cell (the parent cell) and make an exact duplicate by splitting itself into two new identical cells called daughter cells. It is crucial that the process works properly so that the new cells are genetically the same as the original and can do the exact same job. Imagine what problems would result if a mistake occurred in the duplication process and one of the new cells were genetically different. This would mean that the new skin cell healing your cut lip might not function or look like the rest of your skin cells.

To divide evenly, the cell proceeds through the mitosis phase of the cell cycle in a regimented manner. First, in prophase, all the chromosomes are paired up with their like pairs. During metaphase, the pairs line up along the center of the cell. Organelles in the cell, called centrioles, make this happen. Think of these structures as the Spider-man of the cell. The centrioles shoot out protein webbing called spindle fibers. They push and pull the chromosomes until they line up along the equator of the cell, with one of each copy on each side of the equator. During the next phase, anaphase, the centrioles contract the spindle fibers and pull apart the copied chromosomes. They continue to pull until the chromosomes separate, at which time we call it telophase. The last step, cytokinesis, is the actual splitting of the cell. Because of the previous steps, each new cell has a full complement of the genetic information, as each has one copy of the original set of chromosomes. In a human, that is 23 pairs or 46 chromosomes. Remember, this process started with the cell making a copy of the chromosomes so it had two full sets. Now the cell again enters interphase, grows to its original size, rebuilds all the cell parts, and resumes its job in the body of the organism (Figure 4.8).

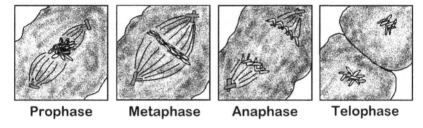

Prophase   Metaphase   Anaphase   Telophase

Figure 4.8: Mitosis Phases

I began this section by telling you this was an orderly process. Despite the best of plans, however, things do not always go as expected. Outside forces, such as sunlight or toxins, can damage the structure of the DNA and/or chromosomes inside a cell. These outside forces are called carcinogens because they can cause cancer cells to grow. The protein p53 is a protein in the cell with the job of stopping the cell cycle if the DNA is damaged by a carcinogen. Under normal circumstances, the level of p53 is increased in damaged cells. If the damage is severe enough, this protein increases to such a level that it kills the cell, effectively causing the cell to commit suicide in an effort to protect the rest of the body from replicating a damaged cell. However, if there is a mistake in the control mechanisms for cell division, this can result in a cancerous growth. A common mistake is a mutation in the gene that codes for the p53 protein. Without this gene functioning properly allowing the cell to make properly-shaped p53 proteins, the cell has no mechanism to recognize its own DNA as being damaged and then prevent the damaged, or mutated, cell from completing its cell cycle. Unchecked duplication of damaged cells ultimately results in a tumor. Another protein, p27, can block the cell cycle from completing for similar reasons. Research indicates that individuals with breast cancer who also have reduced levels of p27 in their cells have lower survival rates. Faulty reproduction of cells might be the most egregious manner in which a cell can be out of homeostasis.

5 – *Respiration and Circulation*

So far in this chapter we have explored mechanisms and strategies cells use to maintain homeostasis. In the previous chapter, we saw regulatory mechanisms for disparate components of ecosystems, communities, and populations. What if we apply the same holistic thinking we used in understanding an ecosystem to understanding how the human body maintains itself, again thinking of our body as an ecosystem?

To do this, we have to understand the organization of the human body system. The smallest unit of organization in your body is the cell. Cells are organized into tissues. There are four kinds of tissues in the human body: epithelial, connective, nervous, and muscle. Tissues are organized into organs, such as heart, lungs, skin, kidney, etc., which are organized into organ systems. The human body systems are the circulatory, digestive, endocrine, immune, integumentary, muscular, nervous, reproductive, respiratory, skeletal, and the urinary and excretory. These systems work together, utilizing cell functions to maintain the homeostasis within your body.

Let's begin with the circulatory system. The blood, blood vessels, and the heart make up the transport system that moves all the necessary molecules throughout your body needed to maintain homeostasis. This is a closed system. Because it is circulatory, it does not really have a beginning and an ending, but we will begin with the heart. The blood in the heart is pumped out through arteries, which are the biggest blood vessels leading away from the heart. Then the arteries divide into smaller blood vessels called arterioles. The arterioles divide into capillaries, where molecules are exchanged between the blood system and the cells. Every cell has at least one capillary connected to it. The capillary is so small that only one blood cell at a time can fit through it. This forces the blood cells to rub against the sides of the capillary. Because of this close contact, molecules being carried by the blood can transport into the cells connected to the capillary (via osmosis, diffusion, facilitated diffusion, or active transport). The capillary walls are very thin, often only one cell thick, which makes the transport as fast and as easy as possible. This is how molecules get into cells from the blood. Conversely, as the blood passes through the capillaries, excess molecules not needed by the cell are transported from the cell into the blood. The capillaries combine into venules, then into veins, and eventually blood returns to the heart.

This is a generalized description of the blood flow. The blood circulation pattern can be thought of like a figure-eight race track, in the configuration of a figure eight with the heart in the middle. Actually, the heart really is like two pumps in one, with each side of the heart responsible for one of the loops of the figure eight. Beginning in the larger, left side of the heart, the larger of the two halves of the heart, the blood is pumped through the first loop carrying oxygen to the cells. It returns to the heart depleted of oxygen and full of carbon dioxide and enters the right side of the heart which then pumps the blood through the smaller loop of the figure eight, which passes through the lungs. The blood releases its carbon dioxide into the lungs and picks up oxygen, returns to the left side of the heart and begins the circuit again to make another "lap" around the circulatory racetrack. As we explore the respiratory, circulatory, and the urinary and excretory systems in the next sections, keep this generalized pattern of blood flow in mind (Figure 4.9).

From this figure-eight, you can quickly see that the respiratory organ system and the circulatory organ system are linked together. Your body utilizes the respiratory system and the circulatory system to get oxygen into your body and ultimately into the cells. This oxygen is needed, remember, to burn the glucose used for energy. In the next section, we will explore how the glucose gets into the cells, but for now, we will focus on how the oxygen enters into the system and the carbon dioxide leaves the system. Recall that carbon dioxide is the waste product of cell respiration.

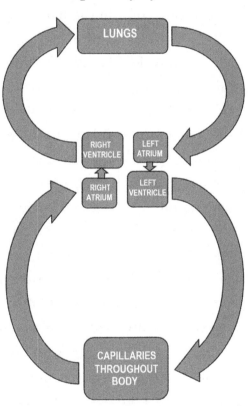

Figure 4.9: Blood Circulation Pattern

Returning to the park, you have finished your popcorn. You pause and take a deep breath, then get to your feet, throw away the empty bag and continue on the path. When you take a deep breath, notice that your chest cavity rises and expands. When this happens, the volume of your lungs is increased and air from the atmosphere pushes its way into your lungs. The air that flows into your lungs is approximately 20 percent oxygen. This oxygen-rich air flows in through your mouth and/or nose, down your trachea, and into your lungs. The lungs branch into smaller and smaller branches, ending in the alveoli, which are little air sacs at the end of each branch—sort of like the leaves on a tree. These alveoli are surrounded by capillaries. The capillaries at this point in the figure-eight loop are carrying blood that is low in concentration of oxygen, but high in concentration of carbon dioxide. The oxygen diffuses into the blood through the alveoli in the lung, and the carbon dioxide diffuses out of the blood into the alveoli, exchanging places with the oxygen. The blood is then pumped back to the heart, where it then is given a boost and pumped out to the body, thus delivering oxygen molecules to the cells that need it to complete cell respiration. When you allow your chest cavity to fall, and thus reduce its volume, this pushes the air back out into the atmosphere and exhales the carbon dioxide that is the waste product from cell respiration out of the body. This connection between the respiratory system and the circulatory system demonstrates how the body's cells are supplied with oxygen and rid of the carbon dioxide waste. Now go back and look at the equation for cell respiration and recall how it shows that oxygen is an input and carbon dioxide is an output in the process of burning food (glucose). This is why you need these two interconnected systems to function at peak efficiency.

6 – Digestion and Circulation

To explore the connection between the respiratory and circulatory systems to the movement of other molecules, including the food burned with the oxygen supplied by the respiratory and circulatory system, we need to add another system—the digestive system. Returning to our park, as you continue your walk, you realize that the popcorn was not enough to satisfy your hunger. Your stomach begins to growl and you recognize the sensation of hunger. You need a more substantial lunch. Your body is telling you that energy reserves stored in your blood and cells from your last meal are now gone. Just as regulating oxygen and carbon dioxide through your respiratory system is crucial to maintaining homeostasis, so is the ingestion and digestion of food. It is time to begin the process of maintaining your energy level by purchasing a hot dog and sports drink from the vendor in the park.

As you eat the hot dog, your body's digestive system kicks into action. First, let's look at the digestion and burning of the hot dog and bun. Digestion begins in the mouth. The first step is, of course, chewing which mechanically breaks the hot dog and bun into smaller components. The smaller the pieces, the greater the surface area, and the easier it is for the digestive chemicals to do their work of breaking down the molecules. The first chemical to act is the protein amylase. Amylase is in the saliva and begins the process of breaking down a carbohydrate called starch, which is made by plants and consists of many simple sugars tied together into long chains. So in our example, amylase begins breaking down the hot dog bun, which is made up mostly of starch.

After you chew the food and amylase begins working on the starch, the process of peristalsis—muscle contractions within the esophagus—helps the food slide down into the stomach. Within the stomach are the next set of digestive juices, secreted by the gastric glands and liver. Pepsin acts on proteins and bile from the liver acts on the fats. It takes about four hours for the stomach to complete its work and pass the rest of the food to the small intestine, where the remainder of digestion occurs (Figure 4.10).

Only the small molecules that make up the carbohydrates, proteins, and lipids in the food are absorbed and used to make specific molecules in your body. The small intestine gets digestive enzymes such as amylase, lipase, protease, peptidase, maltase, lactase, and sucrase from the pancreas and intestinal glands. Enzymes are proteins that cause chemical reactions to occur. Looking at the names of the enzymes, you can probably figure out what type of

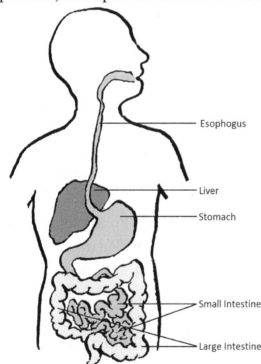

Esophogus

Liver

Stomach

Small Intestine

Large Intestine

Figure 4.10: Digestive System

molecule they help the body to break down. For example, lipase breaks down lipids, protease, proteins, and so on.

Consider now the ingestion and absorption of your sports drink primary ingredient—water. As the sports drink passes through the stomach and then the small intestine where the food is being ingested, the salts and sugars are removed, the water passes through these parts of the digestive system and is not absorbed until reaching the second to last stop in the digestive system—the large intestine. Water diffuses through the walls of the large intestine where it will enter the blood stream to be distributed throughout the body as needed to the body's cells and tissues.

Any substance eaten that is not broken down and absorbed in one of these stops along the way simply passes through as solid waste into the colon for elimination. Everything else ends up in the circulatory system and eventually in the cells where that substance is needed. Can you put the pieces together now and apply the concepts of osmosis, diffusion, facilitated diffusion, and active transport to the digestive process?

Capillaries, the smallest of the blood vessels, surround the digestive system in much the same way they surround the alveoli of the lungs. These capillaries are in close contact with the entire digestive system. As foods are broken down into smaller molecules, one of the transport processes moves the molecules out of the digestive system and into the capillaries. Water molecules diffuse across the membrane and enter the blood stream through the capillaries. Other molecules, such as glucose, fats or proteins, move via facilitated diffusion or active transport depending on their size, electric charge, and the concentration gradient from within the digestive system to the capillaries and into the bloodstream. For example, glucose moves from the small intestine to the body via active transport. After you eat, there is a higher concentration of glucose in the small intestine, but if your body relied on facilitated diffusion only, the glucose would simply diffuse back across the membrane and back into the small intestine as soon as the intestine was empty until there were equal amounts of glucose on both sides, balancing the concentration gradient. This would waste all of the glucose left in the small intestine and you would only absorb half of the available glucose. Therefore, even though glucose can pass through the membrane via facilitated diffusion, active transport is preferred in this situation so all the glucose can be moved which actually creates a situation where the glucose is pumped past the point of equilibrium and against the concentration gradient until there is a very high concentration in the bloodstream and as little left as possible in the small intestine. Once in the bloodstream, the blood transports the molecules (H_2O, CO_2, and nutrients) throughout the body. The larger vessels divide into smaller and smaller vessels until the molecules transport into, and out of the cells, via diffusion, facilitated diffusion, or active transport through the capillaries that surround individual cells and tissues.

7 – Excreting Waste

You have been imaging yourself walking and exploring all morning. You cannot hold it anymore and have to look for a restroom. You find a porta-potty. There certainly is a lot of diffusion of molecules coming from inside! What does your body do with the molecules not needed but that are absorbed into the circulatory system and do not simply pass through as solid waste? These might be toxins from food or from cell processes, or an excess of useful molecules such as water, or simply an excess of cells. Blood is pumped through other organs such as the liver, which filters out toxins like alcohol, and also through the kidney, which maintains the proper level of salts, sugars, and water in the blood through diffusion, osmosis, and active transport.

The kidneys are crucial to maintaining the chemical homeostasis of the blood within the body. For example, many young athletes ingest protein powder in the form of shakes or drinks. This protein powder is an extra dose of amino acids, the building blocks of proteins. Presumably, the thinking is that providing an extra dose of the components of proteins allows the body to build proteins more quickly and therefore add muscle mass more quickly when exercising. However, if eating a healthy and balanced diet, chances are those athletes are getting plenty of amino acids from the food they eat. The excess provided by the protein powder is actually stored as fat or is recognized as extra and then filtered out of the body. The same is true for excess water, salts, sugars, etc. This filtering happens within the excretory system, which is made up of the kidneys, ureters, urinary bladder, and urethra (Figure 4.10).

Within the kidney is an elaborate network of blood vessels, tightly wound around collecting tubules (Figure 4.11). Blood flows into the kidney and spreads throughout a network of capillaries, wrapped around collection tubes. Each network

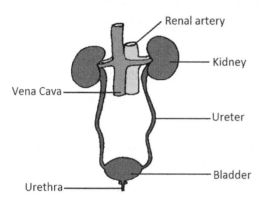

or working unit is called a nephron. One end has a cup-shaped webbing of vessels making up the glomerular capsule. This cup-shaped capsule surrounds the mass of capillaries. The first step of production of urine occurs in the glomerulus.

Figure 4.10 Excretory System

Through four processes—filtration, reabsorption, secretion, and concentration/excretion—the body utilizes osmosis, diffusion, facilitated diffusion, and active transport to regulate the chemistry of the blood. The blood is first filtered in the glomerular capsule where cells, proteins, and other large molecules are removed. Then the blood moves to the proximal convoluted tubule, the loop of Henle, the distal convoluted tubule, and a series of collecting ducts, where it is concentrated and excreted as urine. Along the way, desired molecules are reabsorbed and additional molecules of which the body needs to rid itself are secreted into the forming urine. If the blood contains an excess of water, it diffuses out and one's urine is pale in color. If there is a shortage of water, the urine becomes darker in color, as the water level in your body may drop, but the amount of excess salts, proteins, nitrogen wastes, and sugars will stay the same, making the urine more concentrated. As the blood vessels and the collection duct within the nephron intertwine, the molecules move

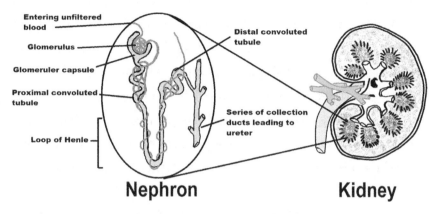

Figure 4.11: Kidney Structure

back and forth until equalized. The body uses active transport to control levels of the molecules it wants to keep, even if they are higher in concentration in the blood than in the urine. This is the case with glucose. The body saves glucose because it is a valuable energy resource, actively transporting glucose mistakenly filtered out of the blood in the kidney back into the blood before the urine enters the ureter on its way to the bladder and out of the body through the urethra. If there is so much glucose in the blood that the kidney cannot actively transport it back into the blood, this is a sign the body is not properly regulating the amount of glucose in its system. In this case, excess glucose will appear in the urine and is the first sign that someone may be suffering from diabetes. Thankfully that did not take too long and you can exit the porta-potty and continue on your walk in the much deserved fresh air.

Homeostasis of the body is the work of all the body's systems. As we have learned, one loop of these systems involves the respiratory system working with the circulatory system to regulate the amount of CO_2 and O_2 in the blood and cells. As these levels change, say due to exercise, breathing rate and pulse rate will change, speeding up or slowing down the movement of these molecules. The other loop of this system involves the circulatory system moving molecules throughout the body between the digestive system and the excretory system. Since food is the energy source for cell respiration, which requires O_2 and produces CO_2 when food is broken down in the cells, this is the connection between the respiratory, circulatory, and ingestion/digestion processes. The waste products and toxins from the cell processes are transported away from the cells by the circulatory system and out of the body either via the respiratory system, in the case of carbon dioxide, or the kidneys as they filter the waste from the blood and eliminate it through urine. These systems, then, work together in a self-contained system to maintain your homeostasis. However, that system is not isolated. It is directly and physically connected to the environment.

Recall the movement of energy in the earth's biosphere. Chapter One introduced the concept of thermodynamics. The second law, in a nutshell, says that all things being equal, the energy in a system will go from ordered to less ordered and become more chaotic. This is called entropy. The higher the entropy, the less ordered the energy is. Natural systems, left alone, always increase in entropy. However, living systems become more organized, not less organized. The individual organisms are the mechanism by which the earth system uses energy to maintain the homeostasis of the ecosystem because that energy drives the function of the organisms that continually recycle the matter.

Close your eyes and imagine the restaurant ball pit again, each ball representing molecules in the atmosphere: oxygen, carbon dioxide, water, and nitrogen. In your mind shrink them small enough so you can no longer see them. Now, with your eyes still closed, relax your breathing. Imagine those molecules rushing into your lungs. Follow the path of an oxygen molecule as it enters your lung, diffuses out of the lung and into the blood. It is pumped throughout the body and diffuses into a cell. Now trace the path of the carbon dioxide out of the cell, through the blood, into your lung and out into the atmosphere, where it is absorbed by a plant and put into glucose molecules. You can eat these molecules along with proteins containing nitrogen. These molecules become part of your cells until you die or maybe simply lose those cells from your outer layer of skin. Decomposers extract the last bit of energy from these molecules and return what remains to the soil. Can you now visualize the physical connections between you and the other components of the ecosystem?

As an individual organism, you take in energy and use it to maintain homeostasis. As the energy flows through the system, living systems *create* order (your functioning body) out of the chaos. This might appear, at first, to defy the laws of physics by becoming more ordered instead of less. However, this is not the case, because all living systems are always *open* systems *within* larger systems. As a whole, the system in which

Video: The Body Systems and Homeostasis
https://vimeo. com/168217214

all livings systems reside, the universe, increases in entropy. The heat dissipated from the body does add to the entropy of the larger system. Individual organisms, however, self-regulate within the context of the larger ecosystem, fulfilling a niche, using the energy to recycle the matter.

Simply by living, your body is participating in the geochemical cycles that maintain the homeostasis of planet Earth. You are merely borrowing from the earth system all of the matter in your body for the duration of your lifetime. You are also adding to the emergent property of that system to help create and maintain Gaia—the earth as a living system. Recognizing that the human body is never isolated is crucial to understanding one's ecological identity. It takes in energy and matter and exchanges it with other members and components of its environment and is therefore physically connected to its larger ecosystem.

Web Resources Home Page
www.exploringbiodiversity.com

Chapter Four Web Resources *http:// www.exploringbiodiversity.com/#!chapter-4-student-resources/ci2d*

Chapter 5: Genetics

What is the role of genetics in our story?

1 – Asexual and Sexual Reproduction

Continuing through our imaginary park, you are surrounded by life. All these living things, including you, carry information within each of its cells. Information is ordered, passed on from parent to offspring, and also incredibly consistent across the species. In the last chapter, we connected an individual organism's functioning to the larger ecosystem, seeing that our connection goes beyond what we easily see at the macro level and involves connections at the cellular level through the exchange of matter. We saw that the individual organism, using energy from the sun, is the mechanism for how the ecosystem recycles matter and creates ordered structure throughout the entire ecosystem. In this chapter, our focus will remain on the individual organism as we look even deeper into the cell. Instead of the use and exchange of matter and energy to maintain homeostasis, this chapter is an exploration of the exchange of information from cell to cell, generation to generation, and how that might inform one's ecological identity.

How much do you look like your parents? How about brothers, sisters, or cousins? Have you ever looked at old pictures of your grandparents as children, and seen yourself in their faces? You are, of course, the product of sexual reproduction and look like others in your family because you inherited traits from your parents, grandparents and so on. With proper genetic testing, this lineage can actually be traced back tens of thousands of years!

Not all organisms reproduce utilizing sexual reproduction. Before we explore how you got your traits such as eye color, hair color, height, intelligence, even possibly a bad temper, we will first explore aspects of reproduction.

Any time you start with a cell containing a full set of genetic information and split it into two cells that are exact copies of the original, that action constitutes *asexual* reproduction. Recall that mitosis is a form of asexual reproduction in eukaryotic organisms to produce new cells—cells coming from cells. Not all cells do even this form of reproduction, however.

Bacteria are prokaryotic cells and do not contain many cell parts. They do have DNA, in the form of a loop called a nucleoid, in place of a nucleus (as in your cells). Because of this simplicity, bacteria generally reproduce through a process similar to mitosis, as discussed in Chapter Four. In bacteria, it is called binary fission. As in mitosis, the bacterium cell copies its DNA and then splits into two new daughter cells. This can occur quite quickly. In the right conditions, say a warm, damp, dishcloth sitting in a kitchen sink, a population of bacteria, called a colony, can produce a new generation every twenty minutes.

Though fast and efficient, binary fission comes with a distinct disadvantage. Genetic diversity is limited because each new set of daughter cells, barring any mutations, inherits an exact copy of the parent DNA. Mutations do happen regularly and because of this, bacteria can evolve quickly, adapting to their environment from one generation to the next, but for the most part, daughter cells are genetically identical to their parent cell.

To avoid this limitation, bacteria have evolved the equivalent of sexual reproduction called conjugation. During conjugation, two bacteria sidle up next to each other and extend a tubular connection called a pilus between them. They trade chunks of smaller loops of DNA called plasmids. This DNA mixes with their larger chunks of DNA in the nucleoid, and the bacteria, each with its recombined DNA, go on their separate ways—each a little genetically different than before the encounter. Now when they complete a cell division by binary fission, the new daughter cells will be different than other daughter cells from the same "grandparent cells," if you will, as they have copied and passed on the newly recombined DNA.

Some types of organisms other than simple prokaryotes can reproduce asexually. It is most prevalent in the fungi and plant kingdoms. Imagine you are getting ready to plant your garden and you are researching potatoes. When you plant a potato, you do not typically plant a seed, which is the result of sexual reproduction; instead, you simply cut up an existing potato, one with the characteristics you want in your potato crop, and put it in the ground. Potatoes can reproduce, like many plants, from a "cutting." Place a portion of a plant in the right conditions and it will grow new cells from the cutting—producing new roots, stems, and leaves. This is called vegetative propagation. The process involved is mitosis, and the resulting plant is genetically identical to the parent plant. It is a clone. In a sense, humans have been cloning some organisms since the agricultural revolution thousands of years ago. A distinct advantage to vegetative propagation is that the resulting plants have known qualities. This is a good thing when gardening. You know exactly what you will get. But when would this not be beneficial? What are the disadvantages to a crop where all of plants have the same genetic make-up?

To answer these questions, let's look for an example in history. In the mid-1800s, the staple crop in Ireland was potatoes. Like today, most farmers planted potatoes using cuttings. Take a potato and cut it up, so each portion has at least a couple of "eyes," and new roots and shoots will grow from those spots. In the small country of Ireland, it came to be that most farmers were growing potatoes that originated from the same parent plant, and they kept using cuttings each year from their previous year's plants. They found a variety that produced a plant with the characteristics they liked, so they used vegetative propagation to produce more and more. It did not take long for almost all the potato plants to be genetically identical, or at least very, very similar.

This was fine until potato blight, a fungus that infected the potato plants, decimated the potato crop of Ireland. If the fungus could invade one individual and kill it, then it had the same effect on all of them, as they were all genetically identical with the same immunities. The blight eliminated almost the entire staple food source, leading to the "potato famine" of the 1840s. Ultimately, a large segment of the Irish population faced a choice of starving or immigrating into other countries. Recall from Chapter Three the effect of changing resources on a population and the choices individuals face when resources are scarce—outcompete others, die, or emigrate. The famine led to a large immigration of Irish citizens into the United States in the second half of the nineteenth century.

Consider another scenario: imagine driving through the countryside in the Midwest region of the United States. As you travel, there are two major crops visible: corn and soybeans. Humans will not eat most of the corn you see growing. Though some of it will be used as an ingredient in processed food, most of it will be fed to the animals we eat or used in industrial products. In most states, ten to fifteen percent of the gasoline powering your car is ethanol, which is alcohol made from plants, often corn. A major concern about the world's corn crops is that genetic diversity is greatly lacking. Modern farmers are not growing corn from cuttings like the Irish potato farmers, but a majority of the corn grown today comes from only a few seed companies. The seed companies have researched the best varieties to cross pollinate, creating plants that have the characteristics the farmers want. However, if you save seeds from those plants and plant them again, they do not grow into the same high-yield corn plants. Therefore, every year, farmers are forced to buy a new set of seeds from the seed company. In fact, the seed companies have patented the corn they have developed and it is therefore illegal for a farmer to save the seeds and try to grow their corn from their own seeds. Putting aside any questions of ethics and legalities, what is the danger of this practice from a biological or ecological standpoint?

If we take a lesson from history, we see that a crop lacking in genetic diversity is subject to collapse. Farmers saving seeds, relying on random sexual reproduction within those plant populations for the next generation of seeds, kept the genetic diversity intact. Now that farmers cannot save seeds and must instead buy them from a few seed producers each year, and the corn has been selectively bred to produce the best kernels, there is the potential for greatly reduced genetic variability. It is now up to the seed companies, as they produce new varieties, to maintain genetic diversity and produce enough different varieties of corn (and other crops) to not replicate the scenario in Ireland that led to major famine.

2 – Chromosomes and Genes

On your imaginary walk, you pass a young woman walking the other direction. She's pushing a double stroller. The two children riding in the stroller are dressed identically and look the same, and you notice the bright red hair and freckles. Reproduction and genetics are quite different for humans than for the organisms discussed in the first section of this chapter. There is no asexual reproduction or other methods reducing genetic diversity in humans. Therefore, most people are not exact copies of anyone else, with the exception of identical twins. However, even the identical twins share an equal mixture of each of their parents' DNA and genes, and therefore a mixture of their parents' traits, and so are not genetic copies of their parents, only each other.

You resulted from sexual reproduction, like all other species in the animal kingdom. It is a slower and less efficient biological process than asexual reproduction, but the benefits far outweigh the difficulties. Chiefly, each generation is genetically different from those in the past. A population of organisms resulting from sexual reproduction is protected against one disease affecting them all equally and wiping out the population. The higher the genetic diversity within a population, the more likely the population is going to have at least some individuals who survive a sudden change in the habitat. We will explore this foundational component to evolution more deeply in the next chapter.

For sexual reproduction to occur, each parent has to donate half of their chromosomes. In the normal state, all the cells in the human body have a full, double set of chromosomes. There are 23 pairs of chromosomes, numbered 1 & 1', 2 & 2', etc., to 22 & 22'. The 23rd pair determines gender and is labeled differently. An X from each parent results in a girl, while an X from the mother (since she only has X's) and a Y from the father results in a boy (Figure 5.1). All cells making up your body, called the somatic cells, contain this full set, with one chromosome of each pair coming from each of your parents. These cells are said to be diploid—"di" meaning two, and "ploid" referring to the chromosomes—and are indicated with the equation 2n, where the "n" refers to the total number of chromosomes.

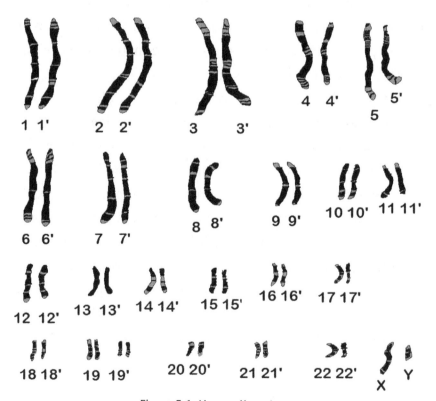

Figure 5.1: Human Karyotype

The process of how you inherited these chromosomes is called meiosis and fertilization. Meiosis produces the sex cells. Notice the term is similar to mitosis. It involves the forming of new cells from a parent cell, as in mitosis, but, instead of producing two identical daughter cells like in mitosis, meiosis results in four genetically unique daughter cells called gametes. Achieving this requires a process that takes twice as long as mitosis with two full cell divisions. The genetic uniqueness comes from a mixing of the chromosomes from each parent. The resulting cells are said to be haploid (1n) as opposed to diploid (2n). Look at the diagram showing the steps of meiosis (Figure 5.2) and then go back and compare and contrast it with the process of mitosis in Chapter Four. As you read through the next sections in this chapter, try to visualize the movement of the chromosomes, and then the genes on those chromosomes, as the cell produces the gametes.

In male animals, the gametes are called sperm. In plants, the gametes are found in pollen. These single cells are haploid and contain half of the chromosome set. For example, a sperm might have the 1 of the 1 & 1' pair, the 2' of the 2 & 2' pair, and so on. If it contains the X of the XY, or 23rd pair, the offspring from that gamete will be a girl. If it contains the Y of the XY, or 23rd pair, the resulting offspring will be a boy. In females of all organisms that have male and female, the gamete is called an egg. The male gamete fertilizes the egg, resulting in a zygote, which is a new cell that is the combination of those two gametes. Within that first cell, the two half-sets of chromosomes combine to make a full set. How the chromosomes separate during meiosis determines which traits you get from your mother and which ones you get from your father.

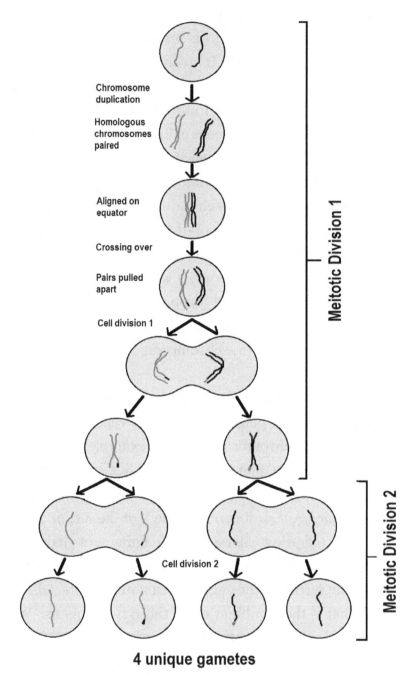

Chromosome
duplication

Homologous
chromosomes
paired

Aligned on
equator

Crossing over

Pairs pulled
apart

Cell division 1

Cell division 2

Meitotic Division 1

Meitotic Division 2

4 unique gametes

Figure 5.2: Meiosis

Early in the study of genetics, scientists thought that the offspring was an even mixing of each parent—sort of like mixing two colors of paint together to get a third color. Gregor Mendel however, discovered a distinct pattern in the passing of traits from parent to offspring. It was not a "blending" at the chromosome level, though sometimes the expression of the genetic information results in traits appearing to "blend." This is particularly true when the trait is a genetic code that results in actual coloration through the production of proteins that are pigments. In these cases, having two versions of a gene results in pigments that do actually mix like two colors of paint.

Mendel, a gardener, was growing pea plants, and noticed there were a number of traits that never mixed. For example, plants either grew tall or short, with no in-between. Seeds were either green or yellow. He also observed the position of the flowers, the shape of the seed pods, and the seed shape and color, among other traits. By fertilizing plants possessing one trait with pollen from plants showing another trait, he discovered that traits were separate and controlled by inherited packets of information that he called genes. The traits that Mendel experimented with were obvious and uncomplicated. As it turned out, a single gene does control each of the traits that he observed. In peas, one single gene causes plant height, another seed color, shape, etc. For most traits in plants and animals, however, this is not the case; tracing most of the pea plants' traits, let alone the traits in an organism like a mammal, is not easy. However, even in humans a single gene does cause a number of traits. Some of these traits are easy to see, such as the ability to roll your tongue, or having a widow's peak, just to name a couple. But what actually is a gene, beyond a packet of genetic information?

3 – Meiosis and the Genetic Code

A gene is a segment of the genetic code on a chromosome; it contains the instructions your cells use to make a protein. The resulting protein causes a trait, such as the ability to produce pigment proteins in your eyes, skin, and hair follicles. Each gene comes in two versions, called alleles; usually one is either dominant or recessive to the other allele.

For ease of illustration, let's look at a human trait that is believed to be caused by a single gene and you noticed on the children in the stroller: freckles. The dominant form of the allele is thought to cause the presence of freckles and the absence of freckles is recessive. This might be a good time to point out that your environment has a big impact on whether or not genes get turned on or to what degree they might be expressed. It is possible for two people with the same capability for producing freckles to end up with different amounts of freckles due to environmental factors such as exposure to the sun.

Recent research has potentially identified the gene that causes freckles on the number 4 chromosome. For ease of visualization, assume the gene is located at the tip of the number 4 chromosome. Now, also assume each chromosome of the number 4 pair (4 and 4') has an allele for this gene. A person could have a dominant on each chromosome, a recessive on each chromosome, or a dominant on one chromosome and a recessive on the other chromosome. A person with one of each is said to be heterozygous for that trait. Two dominant alleles make you homozygous dominant while two recessive alleles, means you are homozygous recessive for that trait. Therefore, thinking back to meiosis, if a person was heterozygous for freckles, they would have a fifty-fifty chance of passing

Video: Meiosis
*https://vimeo.
com/168227507*

on either a dominant or recessive allele. This is because the result of meiosis is that each parent will give only *one* of their chromosome 4 pair (with its corresponding allele for a gene) to the gamete. This chromosome eventually goes to the zygote during fertilization, where it matches up with the other of the pair, resulting in two number 4 chromosomes in the new cell or zygote, one from each parent.

Because this process is so orderly, one can predict the chances of the percentage of offspring that will have specific traits from a set of parents by using what is called a Punnett square (Figure 5.3). This is the pattern that Gregor Mendel worked out using pea plants. A monohybrid cross is used to predict the results of two parents and a single trait. "Mono" means one, and "hybrid" means mixing. When looking at two traits that are on separate chromo-

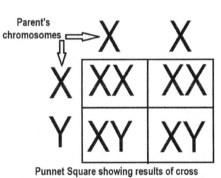

Punnet Square showing results of cross between 23rd chromosome resulting in gender

Figure 5.3: Punnett Square

somes, we use a dihybrid cross. Thinking about how meiosis works, can you picture why two traits on different chromosomes are inherited completely separately from each other?

The principle of independent assortment states that each chromosome pair separates independently during meiosis. For example, whether or not you get the 1 or the 1' chromosome and its corresponding alleles for a trait is independent of whether or not you get the 2 or the 2' chromosome. Because there are 23 pairs of chromosomes and each pair separates independently, there are over 8 million combinations of chromosome pairings each time a gamete is produced. During meiosis, the physical separation of the chromosome pairs from each other is referred to as the law of segregation. Each gamete only gets one of each pair of chromosomes, which segregate from each other early in the process of meiosis. Being able to roll your tongue, having attached earlobes or dimples, even whether or not you are left-handed or right-handed are thought to be traits caused by a single gene. Traits that do not seem to have a pattern of occurring in correlation with one another are very likely caused by genes that are found on separate chromosomes from one another.

Now imagine two traits that often go together such as skin tone and hair color. Either the same gene cause them both or the genes are located on the same chromosome. Therefore, if you get one gene you get the other one next to it on the same chromosome.

The randomness of segregation and assortment is not the only thing that results in unique gametes. A key process that adds to the genetic uniqueness of all individuals is a process called crossing over (Figure 5.4). During meiosis, the chromosomes are duplicated (to make four gametes in the end) and matched up in like pairs (1 with 1, 2 with 2, etc.). During this process, pieces of chromosomes can change places with each other. So

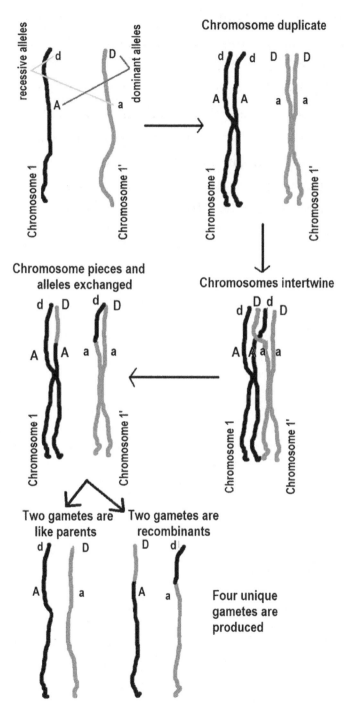

Figure 5.4: Crossing Over

if a 4 has the dominant allele for freckles and the 4' has the recessive, in one of the cells undergoing meiosis these alleles may break off of their respective chromosomes and reattach to the other in the same location. Now they have switched, altering which gamete will inherit the dominant versus the recessive allele. This can happen on every chromosome and in every cell undergoing meiosis. Consequently, there are virtually unlimited combinations that can occur, ensuring that each gamete produced will be different from the others.

4 – The Expression of the Genetic Code

Most traits are not caused by a single gene. Most are polygenic and result from multiple alleles working together. The combination of these alleles—how many dominants or recessives the offspring receives—determines the degree to which each trait is expressed.

Generally, we think that a dominant gene entirely masks a recessive. Therefore, if you have either one or two dominant alleles for freckles, you have the ability to produce freckles. The only way the recessive trait shows is if you have both recessives and no dominant. The expression of a trait and just having the trait in your genetic code, but not showing up on the "outside" is the difference between genotype and phenotype. Genotype is the collection of the actual pair of alleles you have, while phenotype is what actually shows up. Biologists use letters to represent the genes in the genotype. A capitalized first letter of the name of the trait is used to represent the dominant form of the gene, while a small-case letter is used for the recessive allele. For example, the genotype of a person heterozygous for freckles is represented with the letters Ff. A homozygous dominant person is represented with the genotype FF, but both of these individuals would have the same phenotype,—freckles. Only the person with the genotype ff would express, or show, the recessive phenotype and not have freckles.

Not all traits are as simple as either on or off, dominant or recessive. Some are mixtures. A trait caused by a single gene produces a "mixture" in one of two ways. The first is called incomplete dominance. Consider a widow's peak, which is where your hairline dips down into a point in the front (like the Count on Sesame Street). Having the genotype WW produces a pronounced widow's peak. Having the genotype Ww produces less of a widow's peak, and ww produces none at all. This example highlights two things. First, the dominant only partially

Video: Genotype and Phenoype
https://vimeo.com/168333878

covers the recessive, resulting in the intermediate trait. Second, think about all your friends and how many actually have a widow's peak? There are noticeably more people without the distinctive hairline than with, even though the dominant allele causes it. The dominant phenotype is not necessarily more prevalent in a population. It is false to assume that dominant means better or stronger (in terms of a trait surviving and showing up more often in the population). Dominant and recessive only refers to the interaction of the two forms of the gene (the alleles) with each other.

Another way that different forms of a gene can mix is called co-dominance. Human blood type is a single-gene trait with a dominant and a recessive, but in this case there are two equally dominant forms of the dominant allele: A and B. "O" is used to represent the recessive instead of "a" and "b," because while there are two forms of the dominant gene (the A and B allele), there is only one form of the recessive gene (the O allele). Someone with homozygous A phenotype would be AA (getting the dominant A from both parents). In the same way, a homozygous B person would be BB. A person with one of each dominant allele has a third blood type called AB, and someone that is homozygous recessive is OO. Therefore, there are two versions of heterozygous individuals—AO and BO—bringing the total number of possible allele combinations for blood type to six.

The difference between co-dominance and incomplete dominance exists in the interaction of the alleles. Co-dominance results in the dominant completely covering the recessive, no matter which of the dominants is inherited, while an incompletely dominant allele only partially covers the recessive, resulting in a diminished expression of the dominant form of the trait. In co-dominance, a person with both forms of the dominant allele has both of the dominant traits expressed equally in the original form, not a mixture of the two.

By looking at these examples, you can begin to visualize how genes actually work. Having a gene means your body has instructions to make a protein. In human eyes, the ability to produce brown pigment is dominant. Either you get it, or you do not. How dark your eyes are (from black to light green) is a factor of your ability to produce the brown pigment or not. Other factors come into play as well that we will not go into, so eye color is not as simple as "brown" or "not brown." But the ability to produce the protein is an either/or situation. It does not matter to the cell if you get two dominants (BB) or one (Bb) because it only needs the instructions one time to make the brown pigment. If you buy a piece of furniture requiring assembly, would it make a difference if the factory included two copies of the instructions instead of just one? Though if that cell has two recessive (bb) alleles instead, then it does not have any instructions to make the pigment.

Sometimes pigmentation works differently. In some flowers, the ability for red pigment production is incompletely dominant over white. If the allele for red (R) is present, then those cells have the ability to produce the protein that turns them red. However, if it also has the allele for white, then the cell does not produce as much red pigment and some of the white shows through, resulting in pink flowers.

Video: Connecting Meiosis to
Mendelian Genetics
https://vimeo.com/167733208

Video: Crossing Over
https://vimeo.com/168233554

5 – DNA Structure and Genetics

Continuing our walk through the park, you have reached the other side of the park and come upon another playground. From the top of the "castle" tower of the climbing apparatus, a child giggles as he turns round and round sliding down the spiral slide while another child scrambles up the spiral staircase leading to another tower of the "castle." A segment of a molecule called DNA, deoxyribose nucleic acid, makes up the gene. This DNA molecule is shaped like a spiral staircase, or a twisted ladder (Figure 5.5). Phosphates, sub-molecules, and sugars make up

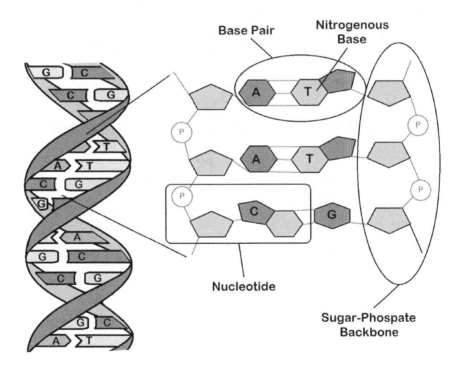

Figure 5.5: DNA Structure

the sides of the ladder. The key to the genetic code, however, is in the rungs of the ladder. Four base pairings of sub-molecules, adenine (A) matched with thymine (T) and guanine (G) matched with cytosine (C), constitute this code. The sequence of bases represents the genetic code and the sequence of these letters (ATCGGTAC...) is what scientists read as the genetic code. Sometimes a gene might be thousands of bases long.

Clearly, this molecule is important to your cell's survival. Yet the DNA molecule is quite fragile. The DNA strand wraps tightly around proteins called histones and is tucked away inside the nucleus of each cell (Figure 5.6). When a cell prepares for division (for either mitosis or meiosis), it first replicates the DNA. DNA replication is quite simple. An enzyme separates DNA like a zipper. As the DNA strand separates, new individual pieces of the DNA are matched with each half. Where a T is exposed, an A is matched with it. As it unwinds, each half is rebuilt by new pieces, so by the time the DNA strand is completely "unzipped," two new identical strands are

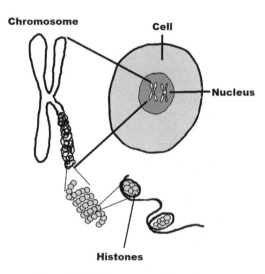

Figure 5.6: CHromosome Structure

formed. Picture yourself unzipping a jacket. As each half of the zipper becomes exposed, imagine individual teeth of zipper material floating in and recombining to match up to each side of the exposed zipper. By the time your jacket is completely unzipped, you would have two new zippers on your jacket. DNA is replicated in a similar manner (Figure 5.7).

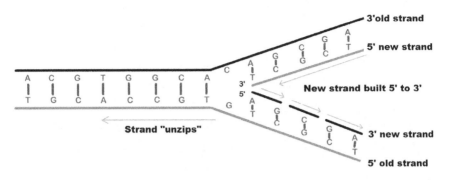

Figure 5.7: DNA Replication

When a cell needs to make a protein, such as pigment in the eye, a two-step process takes place to read the genetic code. First, the cell copies the code of the DNA strand into a single stranded copy called messenger RNA, or mRNA for short. This process, called transcription, happens much like DNA replication described above. The DNA strand with the needed gene is unzipped. New pieces of RNA, instead of DNA, match the letters (A with T, G with C, etc.) and build a strand of RNA with a copy of the DNA code. One difference is that wherever there would be a thymine (T), this base is replaced in RNA with one called uracil (U). This process is called transcription, because it transcribes a copy of the DNA code.

The second step of the process, called translation, begins when the nucleus sends the mRNA out of the nucleus to the ribosome. The ribosome reads each three-letter sequence of letters called a codon (AUC, for example) and matches it up with another piece of RNA called transfer RNA, or tRNA. Each tRNA is specific to each codon, and each codon matches specifically to one kind of amino acid. Amino acids are the individual building blocks of proteins. As each amino acid is placed, the protein is built, piece by piece, according to the sequence of letters from the original DNA code (Figure 5.8).

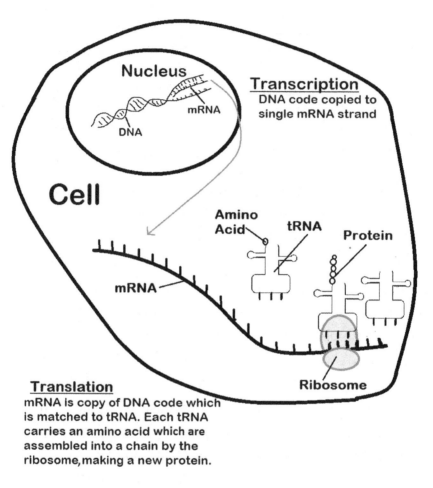

Figure 5.8: Protein Synthesis

Your genetic code is ultimately expressed through these three processes: DNA replication, transcription, and translation (Figure 5.9). When all works properly, the genes you inherited are expressed. However, these complex processes are constantly occurring in your cells and each one involves thousands of "moving parts." Mistakes, or mutations, are bound to occur. Changing just one base pair, or letter (A, G, C, T), alters one codon and could change the sequence of amino acids, which might change the shape of the protein, and therefore affect how the protein works.

Process	Result of Process	Where it Occurs
DNA Replication	Replicates new DNA for a cell preparing for mitosis or meiosis	The nucleus of the cell
Transcription	mRNA is made as transcribed copy of the DNA code to be used later for making a protein.	The nucleus of the cell
Translation	mRNA, with the copy of the genetic code is matched with tRNA carrying specific amino acids – results in protein produc-tion	In the ribosome of the cell, outside of the nucleus.

Figure 5.9 Protein Synthesis Summary

Before we explore the result of mistakes in your chromosomes and which DNA codes cause genetic disorders, it is crucial to point out that the purpose of DNA is *not* to cause diseases. Because of the nature of medical research, we are learning most about the locations of genes on chromosomes related to diseases. For obvious reasons, it is much more important to medical researchers, and more profitable to pharmaceutical companies, to know more about the causes of diseases than eye color.

The first step to finding gene locations is to look at the chromosomes. This is done using a karyotype (figure 5.1), which is a picture of all the chromosomes after they have been removed from a cell. Each base pair on the DNA strand will absorb colored stains differently. So let's say we wanted to discover the gene(s) related to breast cancer. Keep in mind that having a gene

Video: Pro-
tein Synthesis
https://vimeo.
com/168366919

does not mean it will always turn on and cause the expression of that trait. Even so, if we wanted to explore a possible genetic link between breast cancer and genes, we could take karyotypes of a large number of people with breast cancer. Even better would be to compare the chromosomes of people who are related, like siblings, one of whom has breast cancer, and one of whom who does not. By studying the banding patterns on the stained chromosomes and looking for similar patterns among those with the same traits, scientists can begin to identify the location of a trait on a chromosome. It is only the first step, however, and it might be a coincidence that all the individuals with a common disease have a common banding pattern on a chromosome. After identifying the location, the next step is to study the actual sequence of base pairings in that section of DNA and identify the proteins for which it codes. This is the kind of research coming out of the Human Genome Project, which will be explored further shortly.

Another reason to conduct a karyotype is to search for chromosome disorders. For example, it is common for women over the age of forty to have a test done in the early stage of pregnancy called an amniocentesis. During the procedure, the doctor extracts cells from the fluid surrounding the fetus. The chromosomes are removed from the cells, stained, and examined. What the doctor looks for is the full complement (23) of chromosome pairings. The presence of an extra 21st chromosome causes a condition called trisomy-21, or Down syndrome. Extra copies of chromosome 18, chromosome 13, or either of the sex chromosomes can also cause a variety of physical and mental abnormalities in the offspring.

The error in replicating and properly segregating the chromosomes goes back to the early stage of meiosis and the formation of the gametes. As women age, there is a higher likelihood that during gamete formation a mistake called non-disjunction can occur. Sometimes during meiosis, the chromosome pairs do not pull apart properly, so that one of the gametes ends up with both copies of a chromosome. In the case of Down syndrome, or trisomy-21, one gamete receives both copies of the 21st chromosome (both the 21 and 21'), and the other gamete will not get a copy at all. The zygote without the 21st chromosome will not survive. The zygote with the extra chromosome will survive, but having an extra set of those genes leads to a specific type of physical and intellectual abnormalities in the development of the offspring. Conditions like these are called syndromes because they affect multiple genes and result in a multitude of complications. A disease or disorder, by contrast, only affects one gene and therefore only one protein and one trait.

6 – Mutations and the Dynamic Genetic Code

So how does a gene that causes a disease get into an individual or a population? This is the work of a mutation. Every time DNA is replicated or copied for protein synthesis, there is the possibility of a wrong base pair (letter) being put into the sequence. Most of the time, mutations are not noticed by the cells. This is because much of the DNA molecule making up your chromosomes does not actually code for anything, so if a mutation occurs in that section, it never affects proteins that make up the cell. Each amino acid aligns with a codon that consists of three base pairs. However, for many of the amino acids, only the first two bases, or letters, count. The third is actually an extra one. If the mutation occurs on the third letter, changing that portion of the codon doesn't change the amino acid. So again, the mutation isn't noticed by the cell and the majority of mutations to the original DNA strand containing your genes cause no actual changes in the individual. However, sometimes changing even one letter by deleting it, skipping it, substituting a different letter, or adding extra letters changes the amino acid in that protein. This change to the base pair alters the shape of the protein, which changes the functioning of the protein. The protein might not work at all, or it might work in a completely different way than before.

On our walk, you realize that the afternoon sun is high in the sky and hot on your nose and your face. A little sunscreen or hat might have been a good idea. Imagine ultra-violet radiation from the sun striking skin cells during mitosis. The ultra-violet radiation can cause the base pairs to bond in an irregular manner during DNA replication. Adenine (A) could bond with a guanine (G) instead of a thymine (T), for example. This mutation of the DNA strand could change the gene in that section of the DNA, which in turn could change the protein made from that portion of the DNA. This mutation might affect the cell's ability to regulate its own growth and cause it to grow uncontrollably. This is an example of an acquired mutation, because it happens after someone is born and is caused by something from the environment changing the DNA code. This is in contrast to a mutation that a person inherited from their parents.

If the mutation occurs during the formation of the gametes, the mutation is passed on to the next generation, even if it did not affect the person where the mutation originated. Mutations at this level result in new forms of genes. Sometimes these new forms cause diseases; other times they add new traits that are advantageous to a population.

There is a horrible disease called Tay-Sachs that results in a degeneration of mental and physical development, leading to death usually by the age of four. This disease is an autosomal recessive disease, which means it is a recessive allele on an autosomal chromosome (the chromosomes labeled 1 – 22 that cause a majority of your traits). On the 15th chromosome is a gene that codes for a particular enzyme, which is just another name for a protein that your body uses to regulate body functions. In this case, the enzyme in question is one that is supposed to regulate certain fat levels in the blood. Most of the population carries the dominant form of this gene, and therefore most people can properly regulate these fat levels in their blood. In a Tay-Sachs child, there is a mutation, often a deletion or insertion of just one letter (A, T, G, or C) in the sequence for this gene. This change of one of the base pairs, making up the code for this enzyme, results in a recessive version of the allele, which then causes the body to produce an enzyme that doesn't work properly.

Tay-Sachs disease is more common among the population of Ashkenazi Jews. It is thought that this mutation first occurred in the eighth or ninth century. Being an autosomal recessive trait, it requires that the child inherit the mutated gene from both parents. As we know, it's possible to have the recessive version of the gene on one chromosome and the dominant on the other and still be able to make the proper enzyme because of the one correct copy for the enzyme provided by the dominant allele. Therefore, the mutation occurs in the heterozygous parent, who is called a carrier of the recessive allele but doesn't have the disease. That parent has a 50 percent chance of passing on the recessive allele to the child.

Video: Mutations
https://vimeo.
com/168355382

However, since most people are actually homozygous dominant for this gene, the children of carriers usually receive a correct form of the gene from at least one of their parents. If two people who are carriers have a child, they both have a 50 percent chance of passing on the recessive allele that causes the disease. The child has a twenty-five percent chance of getting the recessive allele from both parents, in which case the child will only receive instructions to make the improper form of the enzyme and will inherit the disease. The result of this mutated gene is a change in the shape of the enzyme it produces. Without a properly functioning enzyme, the body cannot regulate the fat levels, leading to a build-up of fats that eventually "choke" the brain.

Tay-Sachs disease cannot be cured because the mistake lies in the genetic code, which is in every cell in the child's body.

Let's consider a hypothetical situation not involving a mutation resulting in death. Assume a mutation occurs during gamete formation that changes the genetic code for the protein controlling brown pigmentation in cells (skin tone, hair color, eye color, etc.). With this new mutation, imagine that the protein produced from that section of DNA causes the cells to produce red pigment instead. Would this change be an advantage or a disadvantage? Maybe neither, but now an individual is carrying a new form of the gene for pigmentation. We have added a new allele to the population. If that individual reproduces, the offspring may inherit this trait. If red pigmentation is a disadvantage, chances are it will eventually be weeded out of the population. If it proves to be an advantage, maybe those individuals will live longer and produce more offspring than other individuals in the population, and then the frequency of this trait will increase. This basic principle of natural selection will be explored further in Chapter Six.

7 – Genetic Technology

Understanding the sequence of the human genome is the work of the Human Genome Project. It is a monumental undertaking with broad medical, social, and political implications. Knowing the location of disease-causing genes might lead to medical breakthroughs, treatments or preventions of those diseases. Keep in mind, however, that once you inherit a gene it is in all of your cells. At one time, scientists thought we could change the genetic code of an individual using a process called gene therapy. The concept of gene therapy is actually quite simple, but virtually impossible. If you can identify the faulty letter sequence in the DNA, then remove it or replace it with the correct nucleotide sequence, you arrive at a cure. This action is quite doable in a single cell, but to cure a disease requires changing the genetic code of *all* the cells in a person's body, which is a much more challenging task.

However, it is possible to recognize the protein error resulting from a gene once the genetic codes are understood. Understanding all the proteins that are the result of your genome is referred to as the proteome. The proteome is to the proteins in your body as the genome is to the genes. Knowing someone's genome tells us all the genes in his or her body; knowing the proteome tells us all of the proteins. The development of the proteome has led to a treatment called protein therapy. It works by identifying a missing protein, due to a missing gene, and providing a medical substitute for that protein. Anyone suffering from diabetes and taking insulin is substituting an artificial source of a protein because his or her body cannot naturally produce it. The more we understand the genes that code for certain proteins, the more specific and precise medicines can become to replace the ones that are faulty or absent.

But what if we could identify the genome of the zygote before it grows into a fetus and change the genetic code at that point, removing the need for medicines or therapies by preventing the disease from developing? What about adding or changing genes not only for disease prevention, but also to engineer children to be smarter, stronger, or better looking? Changing or adding genes at the zygote stage of development is possible because it would only have to be done in a few cells. As those cells divide and differentiate to become the different body parts of the fetus, the modified version of the genome is passed on to each new daughter cell during mitosis. Theoretically, if you change the genome of the first cell after fertilization, then each cell after that would carry the modified genome. For many people this is a scary concept, the possibility of genetically engineering "designer" babies to get the traits the parents want. On the other hand, is it different than providing the best nutrition, the best schools, and the best upbringing for a child? If it was your child, wouldn't you want to do just that? These difficult yet important questions will most likely not be answered by scientists alone, but instead are part of the social and political discussions that need to take place around these topics. As it applies to humans, this type of research is tightly controlled by governmental restrictions as the ethical questions are debated and explored. However, a great deal of genetic engineering is already occurring within the medical and agricultural realms.

It is likely that you ate some corn today. Even if actual corn kernels were not part of your meals, if you ate anything processed—from a box, can or bag—including any beverage other than water, you probably ingested a corn product. Much of the corn grown in the United States is a variety known as Bt corn, which is the result of genetic engineering. You could say that it is a designer crop. Bt stands for *Bacillus thuringiensis*, a bacteria that produces a toxin that repels or kills many types of moths and butterflies. The gene that codes for the protein that is this toxin has been inserted into the corn plant. The corn plants that grow with this new gene will produce the toxin and will be less susceptible to pests. These plants have the code to produce their own insecticide because of the gene inserted from another organism. Their genome has been permanently altered, or engineered. Since insecticides are no longer necessary, the advantage to the farming industry is huge.

"Round-Up Ready" corn is another variety with an inserted gene. This one allows the corn to resist damage from an herbicide that kills all plants. If you have this corn in your field, you can liberally spray the herbicide Round-Up on your crop and eliminate all the weeds that would compete with the desired crop plants without harming the corn.

These corn varieties are just two examples of genetically modified organisms (GMOs). Any organism that has had its genome modified artificially is a GMO. The corn examples are a subgroup of GMOs called transgenic organisms because they have a gene from a different organism inserted into their genome. A great deal of the food we eat has either been modified or is transgenic, which has resulted in higher food production on less land.

Of course, there are arguments against using this technology. What about the insects exposed to this corn and its naturally-occurring toxin that are not pests to the corn, such as monarch butterflies? Will it affect their populations? There is a correlation between a dramatic reduction in monarch butterfly populations and the increased usage of Bt corn. Correlation is not always evidence of causation, however. The population decrease of monarch butterflies could also be the result of climate change, other pesticides, disease, or some other factor not even considered yet. The plight of the monarch butterflies is being actively researched and debated. Another environmental mystery that many people believe to be linked to the use of herbicides and insecticides is colony collapse of bees.

Video: Gentically Modified Organisms *https://vimeo.com/168343686*

Whatever the cause, it is now clear that there is a crisis facing many of the species of bees in the world that we rely upon to pollinate our food crops. I mention the bees and the butterflies to show that the interconnections between all the components of an ecosystem are almost unknowable, but are crucially important.

Will the development of Round-Up ready crops increase the use of herbicides, potentially reducing the biodiversity of plants and insects in areas where it is sprayed? The company that invested in this research now owns this genetically modified strain of corn. It has been successful and most farmers now use it. Will this create a potential repeat of the Irish potato famine? Is it right for a company to own the genome for a staple food? What if this new genome, not the subject to the slow process of evolution, within an ecosystem is released into the ecosystem?

Could a new gene have an effect on the individual or ecosystem similar to an invasive species? Would it have any natural predators to control it?

Genetic engineering has also proliferated in industries other than food production, such as in the medical profession. Much research has explored the human genome and found the causes for genetic disorders that are not directly inherited from your parents. In a sense, everything is genetic.

In 1918, a worldwide influenza pandemic killed millions of people within a short period of time. After about six months, however, the disease essentially disappeared. Doctors did not conquer it with a new medicine or vaccine. The disease ran its course and essentially ran out of people to infect. Those left alive possessed a natural immunity to the disease, and their immune systems fought it off before it killed them. The difference was contained within their genes. So even though influenza is not inherited, the genetic differences within the population accounted for some people surviving and others not. What this tells us is that there is a genetic component to everything we are exposed to in the environment. Understanding these genetic differences, then, could lead to a cure for common diseases.

The insulin used by diabetics is often the result of a genetic engineering process, though not in the same way as GMOs. Scientists insert the gene that produces the insulin protein into bacteria cells. The bacteria incorporate this gene into their genomes and, like tiny biological factories, they produce the desired insulin protein during their own process of reading the DNA and completing protein synthesis. The insulin is collected from the bacteria cells and used as medicine. What if we could identify a gene in a person that was naturally immune to a disease like HIV and could insert that gene into bacteria, use the bacteria to make thousands of copies, and then replicate the protein resulting from the gene? This could be a life-saving medicine created by genetic engineering, specifically transgenic organisms. Is taking medicine produced by transgenic methods in a lab any different than eating a plant containing a gene from bacteria that makes it immune to damaging pests?

These are difficult questions, requiring a great deal of continued public dialogue to address effectively and ethically. The ability of our species to now alter the genomes of organisms is undoubtedly integral to our evolving ecological identity. Meanwhile, research is continuing to produce new varieties of crops we eat and medicines we use every day.

8 - Epigenetics

Are we the way we are because of nature or nurture? The influence of nature (one's genetics) and nurture (one's environment while developing) has been an ongoing debate among biologists, sociologists, and behavioral psychologists since scientists began to learn about genetics. Are you simply a product of your genes, or how much of your behavior and physicality is the result of your environment and upbringing? The answer may lie in an understanding of what is called the epigenome.

Your DNA is made up of very long strands which, in order to fit inside the cell in an organized manner, are wrapped around proteins called histones. The DNA and the histones are covered with chemical tags. These chemical tags make up the epigenome. The epigenome tags cause inactive DNA-carrying genes to wind up tightly so that they cannot be read by the cell. By contrast, active genes are structured by the epigenome to be unwound, making that section of DNA, and therefore the genes on that section, accessible to be read, duplicated, transcribed, and translated into proteins. Remember, it is through these processes that the genes get expressed in your body.

Therefore, nature (your inherited genes) is set for life within the cells of your body. But while the genetic code is not malleable, the epigenome is flexible. The epigenomic tags that cause the DNA to wrap or unwrap are influenced by the outside world (this is the "nurture" part). Chemicals from the environment, or even from within your own body, made from the code of another gene that reacted to a change in the environment can cause the epigenome to relax or tighten up the controls on a section of DNA. This means that another aspect of our ecological identity is that the environment can directly influence whether or not some genes get translated into proteins in your body.

There are many kinds of signals that cells react to that can cause these changes. Cells can come in contact with each other directly, and this contact can trigger a response in one of the cells. Even without direct contact one cell can send a chemical messenger that will activate or deactivate genes in the next cell. Cells can also be activated by hormones, which are specialized proteins designed to trigger a physical or emotional response, like adrenalin does when we're scared. Hormones can be released by one part of the body and quickly sent around to activate genes in other parts of the body. Finally, factors in the outside environment, such as stress and chemicals, can also have an effect on the epigenome.

External signals affect the cell by flipping a switch of sorts. The trigger—an external or internal chemical, contact with another cell, etc.—activates a gene regulatory protein. The gene regulatory protein, because of its shape, interrupts, or spurs on the transcription of the DNA code to the mRNA copy of that code. By doing this, the regulatory protein can cause the activation or deactivation of a gene. This is a short-term impact on gene expression. However, gene regulatory proteins can also have a long-term impact by changing the epigenetic tags on the DNA strand. In this way, they change how tightly wound or unwound the DNA strand is, thus changing more permanently whether that section of DNA can be translated into a protein or not. This change is "remembered" by the cell until a new regulatory protein, chemical stimulant, or other trigger changes the epigenome again.

Therefore, though the genetic code remains the same, whether or not a gene gets turned on or off can be changed by the environment, which is something you can control. As an individual you can control the chemicals your body is exposed to or consumes; you can control the types of stressful situations you put yourself in and even how you respond emotionally to that stress. Controlling these factors can actually have an impact on the expression of your traits. Therefore, the more we learn about the genetic code, epigenome, and what can trigger the wrapping or unwrapping of the DNA and thus the reading of the code to be expressed within the cells, the more we may be able to understand the effect our environment has on the expression of our traits within our cells

If every one of your cells carries the exact same genetic code, how is it that your body is made up of many different kinds of cells that do a wide variety of jobs? Somehow your body ends up as cells which can produce different kinds of proteins. Even though all coming from the same genetic code, your body is made up of cells that are highly specialized and different from one another. Beyond a similar cell membrane, the red blood cells carrying oxygen throughout your body do not have much in common structurally with the heart muscle cells in your ventricle. Yet, somehow your body knows which cells to make where. This mysterious ability is linked to the concepts of the epigenome and that not all genes are turned on or off in all of our cells.

When an egg is fertilized and the zygote begins to develop, all of the cells are initially the same. They are stem cells, which means they have the capability to form into any kind of cell. Early in the development of the fetus, however, the epigenome begins to guide the differentiation of the cells into specialized form and function. As the embryo forms, multiple signals are received by the epigenome during the days, weeks, and months of development. These signals cause changes in the epigenome and therefore influence which genes will be expressed as that specific cell develops. The epigenome "remembers" these changes, so from that point forward only the genes necessary for that cell are turned on, and the rest are deactivated. In the end, a differentiated cell may only have ten to twenty

Video: Epi-
genetics
*https://vimeo.
com/168241423*

percent of its genetic code activated. That is, until environmental factors trigger changes in the gene expression. As a cell grows and divides, either through mitosis to make new cells (like when you need to replace cells to heal an injury) or through meiosis to make gametes, the epigenome is replicated along with the DNA code. Therefore, if at some point in your life a cell's epigenome is affected by the environment, triggering the activation or inactivation of a particular gene, this information is passed on to the daughter cells. So the cells learn their role, and then pass on that information to the next generation of cells.

As much as we like to think that our bodies can be understood in isolation, it is clear that this is simply not the case. Not only are we physically connected to our environment by playing a crucial role in the exchange of matter and energy as seen in chapters one, two, and four, it is now clear that the ecological identity of our species, and therefore of each individual, also includes the passing of genetic information from one generation to the next. Not only is this passing on of genetic information a part of our ecological identity, but the environment itself can change the genetic information in our cells that we pass on. It is clear that we are truly living *in* a biological world—or maybe it would be more accurate to say *within* a biological world. One thing that we can conclude from the new science of epigenetics is that we can no longer consider that we can remove ourselves from the influences of the natural world. Epigenetics shows us that we are inextricably linked to the natural world within which we reside.

Web Resources Home Page
www.exploringbiodiversity.com

Chapter Five Web Resources
http://www.exploringbiodiversity.
com/#!chapter-5-student-resources/crj6

Chapter 6: Evolution and Biodiversity

What is the evolutionary history of our biodiverse planet?

1 – The Beginning of our Story

Up until now, we've been imagining ourselves in a city park, exploring all the ways we interact with that ecosystem, from consuming energy and transforming molecules to how the environment might change our genetic code. Now let's switch to an entirely different scene in an entirely different time. The location of this new visualization is somewhere on the eastern half of the African continent. The temperature is hot, approaching 90 degrees Fahrenheit; dry winds bring smells from the west. An adult female walks among a small group—possibly a family group. She is like you in many ways. However, unlike you, she is only a little over four feet tall. Her arms are much longer in proportion to her torso than your arms. The lower portion of her jaw protrudes much further out than the jaw you inherited from your parents, and she has a noticeably absent forehead. As she walks, she scans the horizon for potential dangers, just as you would in this same situation. Walking

Video: The Earth is an Apple
https://vimeo. com/167732398

199

alongside her is her child. We know this scene occurred because paleontologist Mary Leakey discovered footprints left by two human ancestors, some 3.6 million years ago in Laetoli, Tanzania. The footprints, discovered in 1976, were most likely made by *Australopithecus afarensis* (Figure 6.1). Their path was preserved in wet volcanic ash that hardened like cement. A few years prior to Leakey's discovery, Donald Johanson discovered a remarkably complete skeleton of *A. afarensis* in Ethiopia. It is one of the most famous hominid fossil finds and

was named "Lucy" by Johanson's team. What relevance might this 3.6 million-year-old story have for you today, as you walk down a busy city sidewalk or through a carefully constructed park?

What is my history? Where do I come from? What is my role? How does my history help determine who I am today?

Figure 6.1: *Australopithecus afarensis*

Who are we as a species? These are all *big* questions and worthy of exploration. Understanding one's ecological identity requires looking at our species' evolutionary history.

Our species, *Homo sapiens,* is 150,000 to 200,000 years old. In terms of geological time and the evolution of biological life, that is blip in the history of life on Earth. An ever-increasing list of primates that are direct ancestors to our species, or "cousins" to our species, fills the human family tree (Figure 6.2). The development of our family tree is (excuse the pun)

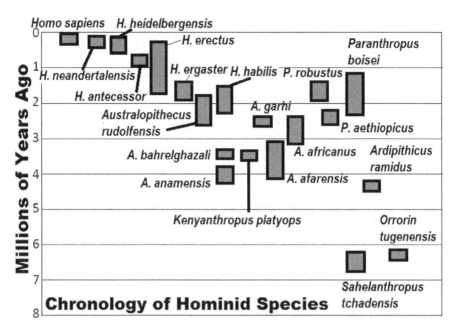

Figure 6.2: Hominid Evolution

an evolving story, perhaps because it is *our own* story. Not surprisingly, there is a great deal of scientific interest in, and a great deal of research funding for, digging up the past. Each new fossil and archeological find adds a new piece to the puzzle and a new chapter to the story; sometimes it even changes the story. Because of this, scientists understand the overall story quite well—though the details continue to change. When *Homo sapiens* evolved in Africa 150,000 to 200,000 years ago, there were other forms of humans as well, and many had come before our species, but for some reason, our species is the sole survivor at the tip of one of the branches of the human family tree. Our closest biological cousins living today are the chimpanzees. Make no mistake, chimpanzees are *not* our direct ancestors, but they do share a common ancestry with us that dates back to about 7 million years ago. Our common ancestor is probably a creature that would be in the *australopithecine* family like Lucy, or possibly that species is the missing link leading to one branch of organisms from which our species evolved and another branch from which the bonobo chimpanzee evolved. It is tempting to think of our species as more evolved than a species like the bonobo chimp, but it is important to realize that both species have equal amounts of evolutionary time behind them and are equally evolved to fit their ecosystems. While our brains continued to evolve and increase in size and capacity, it would be wrong to assume the same thing about all of our traits as compared to the chimp. For example, an anthropologist at Stoney Brook University, Dr. Sergio Almecija, recently published findings that support the idea that the human hand has evolved much less than the chimpanzee hand in the last few million years, meaning that, in this one trait anyway, the chimpanzee is *more* evolved than we are.

Before we focus more deeply on our family tree, let's recall that our purpose is to understand our history within the context of the biosphere. Therefore, we must first explore the evolution of all life on the planet—how it might have originated, how evolution works, how genetics are a part of evolution, and how the ecosystem and the individual evolve together—before we can return to the question of our evolutionary history.

2 – The Origin of Life on Earth

Leaving behind the visualization of the African savanna we can return to the city park. Notice a massive rock of basalt in the park. It is big enough for individuals to climb on top of and sit in the warmth of the sun. This igneous rock might be as old as 3 billion years. The earth is even older. It was much different when life first evolved than it is today with its current abundance of biodiversity: a wide variety of plants, animals, bacteria, fungi, and single-celled organisms. Of course, "old" is relative. Geological evidence suggests the earth is around 4.6 billion years old. In contrast, astronomical evidence has led astronomers to conclude that the universe is approximately 13.7 billion years old. The earth's early atmosphere was much different than it is today. As we know, the four main components of all living things are carbon (C), hydrogen (H), oxygen (O), and nitrogen (N). The early atmosphere was composed of volcanic gases containing these elements in molecules such as methane (CH_4), carbon dioxide (CO_2), nitrogen gas (N_2), ammonia (NH_3), hydrogen gas (H_2), and water vapor (H_2O).

In 1952 at the University of Chicago, Stanley Miller and Harold Urey, tested a hypothesis that if they combined the elements in the atmosphere with energy in the form of lightening and heat, the first molecules of life could spontaneously form (Figure 6.3). To test their hypothesis they created a model of the early atmosphere with gas chambers consisting of gases thought to be present at the time. They added heat to boil the water, which simulated volcanic activity in oceans, and an electric spark to simulate lightning. Amino acids (the building blocks of proteins) were formed. In a May, 1953, article, "A Production of Amino Acids under Possible Primitive Earth Conditions," Miller identified five amino acids resulting from his experiment. In 2009, a different experiment showed how nature could have spontaneously assembled the essential sugars and nucleotides to form the first kind of genetic material—RNA. But it is a long road from those molecules to a living thing.

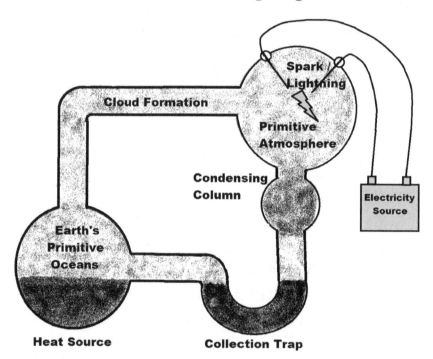

Figure 6.3 Millery-Urey Experiment

For something to be alive it has to be autopoietic, or self-regulating. This concept consists of two interacting principles. First, self-regulating means it can maintain its own order and organization. This happens with living systems when all the combined cell functions are utilized to maintain homeostasis as previously discussed. The second principle is the ability for self-replication. This latter trait is at the center of the debate about classification of viruses as living or non-living, as they have many characteristics of life but do not replicate on their own. Despite this limitation, even viruses can evolve because they have a mechanism to store coded information. The ability to store coded information, copy it, and pass it on to the next generation is essential to evolution and is also, as it turns out, a key characteristic of living systems. No living system can avoid the inevitability of evolution, whether it's a multi-celled species of whale, a colony of single-cell protozoans living together, or a single-celled bacterium.

3 - The First Cells and Gaia

There are two hypotheses as to how the first "cells" evolved. The first hypothesis is that life did not initially form within enclosed cell membranes, but instead formed as molecules or crystals that could self-replicate, showing the first signs of autopoiesis. The second hypothesis involves pre-cells, called coacervates, which are clusters of proteins or protein-like substances held together in small droplets within a surrounding liquid. Within these droplets, simple organic molecules may have combined to form larger complex organic molecules, marking the beginning of autopoietic life functions.

Whatever the form of the first cells, it is likely they were heterotrophic and not autotrophic. Photosynthesis requires much more complexity in cell processes and molecules (mainly chlorophyll) than were present four billion years ago. So it is more likely that primitive cells absorbed and digested organic compounds for food. As the cells evolved, some must have developed the ability to use visible light from the sun as a source of making their own food.

These first cells were most likely primitive prokaryotes lacking a nucleus or other cell parts, and probably were also anaerobic, meaning they lived in areas without oxygen. The early atmosphere was lacking in free oxygen. Lynn Margulis, from the University of Massachusetts, proposed that smaller prokaryotes began living inside of larger prokaryotes to form the first cells with interior parts. The genetics of these organisms supports this hypothesis, as even in the cells of humans today, mitochondria have their own separate DNA that is much more similar to prokaryotic DNA than eukaryotic DNA. Additionally, the chloroplasts within plants contain their own prokaryotic-like DNA separate from the plant's eukaryotic DNA.

This suggestion that smaller prokaryotes, with advanced abilities to metabolize food for energy (like mitochondria inside modern eukaryotic cells) and converting sunlight to food (like chloroplasts in modern plant cells), were absorbed into larger bacteria and formed a symbiotic relationship is called the endosymbiosis hypothesis. When modern eukaryotic cells replicate (as in meiosis), they pass on a mixture of their DNA within the chromosomes. Mitochondria, on the other hand, pass on their mitochondrial DNA, or mtDNA, as an exact copy (excepting any mutations) to the next generation of cells. This crucial evidence for the endosymbiosis hypothesis also has been an important piece of genetic evidence in studying human evolution and making predictions about the age of our own species, *Homo sapiens*. Amazingly, researchers now can predict the ancestry of this mtDNA in the human line back about 100,000 years!

Video: Endosymbiosis Hypothesis
https://vimeo.com/168781846

It is important to understand the relationship between the biological life on Earth and the physical or geological earth. Many people envision life evolving on the earth as a reaction to environmental changes. The real story is more complex. Life on Earth and the physical planet itself are interconnected; in many ways, Earth itself shows characteristics of being autopoietic and regulating its own homeostasis. When the first life on Earth began to do photosynthesis, this added oxygen to the atmosphere. As oxygen was added, the amount of carbon dioxide was reduced. As carbon dioxide levels fell, the atmosphere captured less heat and the temperature of the earth changed. The living things on the planet helped regulate the atmospheric gasses, which in turn regulated the planet's temperature. The geology of the planet itself, such as the amount of reflective ice versus heat-absorbing dark rock, also contributes to the regulation of the planet's temperature. It is the complex interaction of all the biotic and abiotic parts of the ecosystem working to maintain a constant temperature that James Lovelock named the Gaia Hypothesis. The ramification of such an idea is that all living things (including humans) are connected as functioning components of a larger system. The evolution of life changed the earth, which then further affected how life evolved on Earth.

Building on the idea that all life is connected gives us a good paradigm for understanding connection through our shared evolutionary history. Foundational to the study of biology is the premise that all life on the planet evolved from a single common ancestral cell. The supporting evidence is both physical and genetic. The oldest fossils of bacterial cells are approximately three billion years old. But the most important evidence for this connection is in the cells of all living things today. All life on planet Earth carries its genetic information in the same manner—DNA utilizing four base pairs: adenine, guanine, cytosine, and thymine.

Beyond similarities of pattern and function at the genetic level as evidence for evolution are also similarities of pattern and function at the protein level and morphology (body structure) level. It makes sense that if some of the DNA is the same, then many of the genes will be the same and many of the proteins produced by those genes will also be the same. This hypothesis is easy to test by comparing the proteins in a variety of organisms for similarities and differences. Indeed, the case exists that proteins produced by one animal can be readily used in another animal because they come from the same gene, showing common ancestry. The use of bovine insulin to treat people with diabetes (the lack of ability to produce insulin to regulate blood sugar) is an example of an interchangeable protein.

On a macro level, we can also see similarity of pattern and function. The skeletal and embryonic structures of all vertebrates share many commonalities. Though each species has evolved specialized adaptations to survive in particular niches in their specific habitats, they share the same basic skeletal structure.

Despite this commonality, throughout the history of life, millions of different species have evolved (and then gone extinct) as the environment has changed, continents have moved, climates have changed, etc. The great variety and diversity of life we see all around us when we're walking a city street or in a park comes from a common ancestor. Your ecological identity, whether you have recognized it or not, includes that you are a part of that biodiversity.

4 – The Mechanism of Evolution

Walking further through the imaginary park now imagine seeing and hearing a greater variety of birds. In addition to the diversity of songbirds, there are now larger birds common to the ocean, such as seagulls adding to that diversity. To understand the diversity of life that has evolved, we must understand the species concept. Each different kind of living thing is called a species. Biologists came up with the concept of species before they understood genetics. It seems logical now to use the genetic code as the basis for classifying organisms. However, the definition we still use is the one created a couple hundred years ago: two organisms are different species if they cannot reproduce and produce viable offspring. Viable means that the offspring resulting from the mating can live a full life and reproduce successfully as well. That last part is the important part, as the ability to pass on genes to the next generation is at the heart of the biological definition of "surviving."

From the perspective of the species, merely surviving as an individual does not aid the survival and evolution of the species as a whole. The individual must pass on traits to the next generation. When individuals within a population can no longer reproduce successfully, then speciation has occurred. Speciation is a crucial step in evolution, as it represents the point in the history of a species when it branches off from a common ancestor shared with another species. Where each species branches off from the others is often represented with a graphic called a phylogenetic tree (see the human family tree in Figure 6.2 above). The reproductive isolation that causes the branching might be the result of behavioral differences (like being nocturnal versus diurnal), genetic mutations that make two individuals genetically incompatible, or geologic separation, such as a river or mountain range physically separating two populations of a species. Separated populations will continue to evolve to match their specific habitats. If those habitats are different, and the now-separated populations do not have a means to share genes, they will continue to evolve over time, quite possibly becoming significantly different from one another (Figure 6.4). This is called speciation.

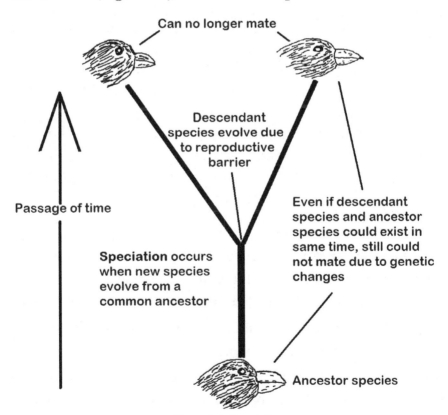

Figure 6.4: Speciation

Charles Darwin was not the first person to introduce the idea of evolution, but he was the one who proposed a specific theoretical mechanism as to how it might actually occur. In 1831, Darwin hired on as the naturalist on an expedition of the H.M.S Beagle for a five-year voyage, which took him across the entire globe. Although he collected samples of animals from around the world, the most famous and important ones came from the Galapagos Islands off the west coast of South America. There he saw evidence of many organisms that had gone extinct, as well as evidence of organisms that were out of place, such as sea creatures high above the ocean in the Andes Mountains. These findings suggested to him that the earth was changing; he hypothesized that life on the earth must change as well or go extinct.

He noticed the patterns of unity of structure and function previously discussed. On the Galapagos Islands, he found many species similar to mainland counterparts, yet uniquely adapted to specific habitats on the various islands. Most famously, he catalogued 13 species of ground finches. This was not a lightning bolt of revelation for Darwin (or like an apple falling from a tree like the mythology surrounding Newton), but was the result of many painstaking years of continued study

Video: The Evolutionary History of Darwin's Finches
https://vimeo. com/167732547

and upon returning from his voyage, he continued to study variations within a species and variations among similar species. He studied barnacles, domesticated animals and pigeons, and continued to hypothesize about the origin of those 13 different species of ground finches he collected on the Galapagos Islands (Figure 6.5).

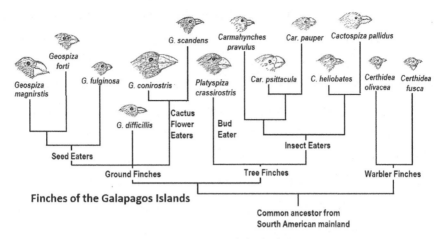

Figure 6.5: Darwin's Finches

In 1859, Darwin finally published his book that would change the study of biology forever. *On the Origin of Species* is arguably the most important publication in the history of the biological sciences and possibly in all of science. In his book, Darwin used evidence of pattern of structure and function to explain how species evolve from one another and ultimately from a common ancestor. This theory is possibly the most challenged and studied theory in the history of science. Since its publication, much more has been learned about the specifics of the mechanism of evolution (particularly genetics), and the theory of evolution by means of natural selection has been observed and supported by experiment after experiment. Laboratory and "natural experiments" where evolution is observed in nature continue to support a shockingly simple process of evolution by natural selection.

There are numerous, real-world observations of natural selection. One affecting people every day is the evolution of bacteria and viruses. Humans continue to get colds and the flu because the viruses continue to evolve as our immune systems continue to learn how to destroy the viruses. Viruses are of course very tiny, but we now have the technology to map their DNA and see if it has changed from one year to the next. The evidence of its evolution is much easier to see. New vaccines are needed every year as the old ones are rendered ineffective by the evolving viral genetic codes in our continuing evolutionary arms race with the microscopic organisms that make us sick.

We should be even more concerned about the evolution of bacteria. Prior to the 1940s there were no medicines for bacterial infections. If your body could fight it off you would live. If not, then you would die. The bacteria and their human hosts were co-evolving, though at a relatively slow rate. Enter the wonder drug penicillin in the 1940s. Recall that penicillin is a toxin produced by fungi as a natural defense (for the fungus) against bacterial infections. When this property was recognized and capitalized upon to make the miraculous, life-saving drug, the relationship between bacteria and humans changed. Prior to penicillin, more soldiers died from bacterial infections, than at the hands of their enemies. World War II was the first war where this was not the case, and it is due to the advent of penicillin.

However, there is a down side to this powerful drug. Its success has led to overuse. Because it worked so well with infection, doctors began prescribing penicillin (or variations of penicillin) for nearly all common illnesses, and that trend has continued. Much of the time, the patient is not sick because of a bacterial infection, but from a virus. Antibiotics do not help the body fight viruses, but the antibiotic does kill some of the bacteria living in the person, even if it is not making them sick. Though many of the bacteria killed have symbiotic relationships with the host, the real danger from this over prescription of antibiotics lies in how the antibiotics affect the bacteria that do make us sick.

Antibiotics work really well, except that they do not kill *every* last bacteria residing in a person; a few individual bacterial cells always survive. Those that survive do so because they have a genetic code making them different enough from the rest of the population that they are resistant to the antibiotic. All the bacteria that grow through conjugation and binary fission from this surviving individual are now also resistant. Prior to the antibiotic's use, this variant may have been a minority in the population. However, it is not difficult to see how this variant quickly becomes the majority of a bacteria population. Because of this, new antibiotics are needed regularly to keep up with our arms race with the bacteria.

But those are simple bacteria, you might be thinking to yourself. How can this evolutionary process apply to more complex animals? The peppered moths of England are classic examples of natural selection. These moths got their name because of their coloring: mostly white with black specks. They evolved to blend in well with the white, speckled bark of the trees in their preferred habitat, providing camouflage from predator birds. Beginning with the Industrial Revolution of the mid- to late-1800s and the advent of coal as a widely-used fuel source, the introduction of voluminous amounts of soot into the air slowly turned the trees darker and darker over the years. As the trees turned black, the individual moths with darker pigmentation came to dominate the moth population, just like the antibiotic-resistant bacteria come to dominate a bacteria population. As the air pollution problem in England was addressed, the trees returned to their original color, and soon the population of peppered moths shifted back to more individuals with lighter coloration (Figure 6.6). So what was occurring inside the DNA of these organisms to cause the population to change?

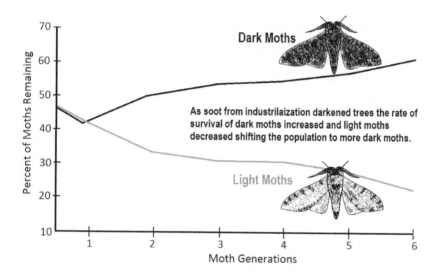

Figure 6.6: Peppered Moths

This type of change in response to environment is the reason Darwin named his theory "evolution by natural selection." *Nature* is selecting. It can be broken down into amazingly simple steps, and when you take each step one at a time, it is easy to visualize and comprehend the logic and simplicity of this process that so often seems incomprehensible to many.

First, an organism produces more offspring than can survive because the best way to ensure the survival of your genes is to reproduce as much as possible. The more offspring you produce, the more likely your genes will survive into the next generation. In many ways, it is easier to think of the genes themselves as individual organisms "selfishly" (as Richard Dawkins put it) working to pass on as many copies of themselves to the next generation as possible. The general rule is that the *less* a species takes care of its young or the *lower* it is on the food chain, the more offspring it produces. High numbers of offspring that are vulnerable and/or highly preyed upon are needed to ensure that at least some survive to the next generation. Think about how many offspring a fly produces in its lifetime as compared to a grizzly bear.

Second, the individuals produced are all unique. There is variation between them. To the untrained eye, one seagull flying over the park looks identical to the other. But by looking more closely, you can find a great deal of variation in physical attributes (such as beak size) or intellectual attributes (such as the ability to navigate migratory routes). Unless two individuals are identical twins, they will not be the same. Therefore, some will have advantageous traits over the others.

Third, the offspring will compete with each other. We often think of competition in nature as something between different species, such as in predator-prey relationships. But the lynx that hunts the snowshoe hare is not competing with the snowshoe hare. The lynx is competing with other lynx to catch the most food, grow to be the strongest, and find the most mates. Conversely, the individual snowshoe hares are competing with one another to get away from the lynx, find appropriate hiding places, and find the most mates. Those with the best traits will survive the most.

Lastly, those that do survive will pass on their genes to the next generation. The offspring of the next generation will have many of the traits of their parents. The traits that are advantageous (say, camouflage or large feet to stay on top of the snow) will be found more frequently in the population than those traits that don't aid survival of the individuals. As the frequency of these genes increases in the population, the genetic makeup of that species changes as a whole to reflect these traits. Conversely, the traits that inhibit survival are weeded out as the individuals with those traits have a lower survival rate within that population.

In addition to nature "selecting" which traits are advantageous and which are not, and therefore selecting which individuals survive and which do not, there is another factor that is equally influential on which genes and traits survive and proliferate within a population. Almost all organisms that participate in sexual reproduction are subject to "sexual" selection. This is the process by which the act of choosing mates influences which traits are selected for and which are weeded out of the population.

Sexual selection and natural selection are often at odds with each other, pushing and pulling the evolution of a species in opposite directions. For example, let's consider the evolution of the cardinal. The male of the species is bright red, while the female is dull in color. Considering that survival is about making sure that the maximum number of your genes are passed on to the next generation, it is the genetic goal of both the male and female to have, and keep alive, as many offspring as possible. To do this, the male wants to attract the mate and live as long as possible to mate as many times as possible. Each spring, the male cardinals begin staking out their territory by finding a good perch and nesting site and starting to sing. Remember, this is the bird that sings, "looook heeere, looook heeere, here I am, here I am." Keep in mind these are bright red birds, singing loudly while sitting in a tree just as the leaves are leafing out—it is safe to say that they are not hiding or well- camouflaged. In fact, the opposite is true. The traits that make the males more visible to females are completely opposite from those that would help them hide from predators. The male cardinal compromises the benefits of hiding from predators in favor of attracting a mate. Meanwhile, the female of the species doesn't say a peep—literally—and is dull in color. As is often the case with sexual selection, the male has evolved characteristics that demonstrate his genetic fitness to the female in order to increase the chances that she will notice him over the neighboring male cardinal.

It is easy (and dangerous) to begin thinking about this process from an anthropomorphic paradigm and to think of it like *we* would when we make choices. It would be inappropriate to think of the male purposefully choosing a louder call and brighter coloration and the female consciously thinking, "Wow, he's got bright feathers and can sing really loud—he must be quite something!" When we begin to think that way the process of evolution seems too complex and intentional to be the result of random mutations modifying the genome. The reality is actually much simpler. In the case of the male cardinal, the tug and pull of natural selection and sexual selection keep him bright enough and loud enough to attract a mate, but because it also makes him more susceptible to predators, he must also have the speed, alertness, and cunning to avoid predators if he is to survive long enough to mate. It makes sense, then, that only the males that possess all these traits—those that aid survival and those that increase visibility—succeed in mating and get to pass on their genes. The key to understanding the evolution of seemingly competing interests, complex behaviors and structures, and the time it takes for these changes to occur is to continue asking the simple question, "Does this trait in any form help the individual survive and produce more offspring?"

Consider this: what is the point of half of a wing? How is it possible that insects evolved wings a little at a time? Wouldn't a wing have to spontaneously appear in one generation for it to be enough of an advantage to help an individual survive better than its competitors? James Marden and Melissa Kramer, from the University of Pennsylvania, researched this question. Their hypothesis was that even the merest nub of a wing might aid the survival of a species like mayflies or stoneflies, which are large flying insects that live as larvae in freshwater. After they complete their metamorphosis into adult flies, they emerge from the water and fly away. The faster they can get off the surface of the water, the less likely they are to be eaten by a surface-feeding fish. But how could a full wing evolve? It would take possibly thousands of generations to go from no wing to a full wing. This conundrum is always present when trying to understand the evolution of complex structures like a wing or an eye.

To test their hypothesis, Marden and Kramer filmed stoneflies on the surface of the water. They clipped the wings incrementally smaller and each time filmed the stonefly skimming across the water. When the wings were clipped too small to allow the insects to escape the surface tension of the water and attain flight, they were still able to use the wings to help them skim across the water faster than if they did not have even the nub of a wing. The larger the nub, the faster the insect could skim across the water to escape to the safety of the shore of the pond, lake, or stream from which it hatched. Looking at the process incrementally like this, it is easy to comprehend that the individuals with even the smallest nub of a wing would have a higher survival rate than those with none, which would then lead to more individuals in each generation having the adaptation. If in each generation there existed a range of nub-wing sizes, and if even the slightest increase was an advantage to survival, each generation would see more individuals with larger wings. With enough generations, the wings would eventually have sufficient surface area to support flight.

So where did the first nub of a wing come from? It could have been a random mutation producing a new structure, but this is highly unlikely. Evolution does not work that quickly or conveniently to simply produce something completely new in one generation—even the nub of a wing that could then slowly evolve into a full-fledged wing. More likely, the first wing-nub was actually a structure that was originally evolved for another purpose. In the case of the stonefly it is conceivable that the nub began as a small, paddle-like structure for moving water around the abdomen to increase oxygen absorption before the insect's emergence from the water. The process of natural selection is often a story of an adaptation that evolved for one purpose also having an unexpected benefit for an entirely new behavior or trait. Gill-like paddles on the side of an insect's abdomen that, once out of the water, rapidly move back and forth, aid those individuals' survival. A few insects in the next generation have even larger ones and survive at an even higher rate. After enough time that adaptation may no longer be useful for the original purpose, but instead is useful for an entirely new reason. As long as it aids survival, the adaptation will remain and possibly even increase or be refined and improved upon over time. Fossil evidence from approximately 300 hundred million years ago supports the hypothesis that ancestral stoneflies and mayflies had partial wings or folded-over wings that were too small to allow the insects to fly, but that could have, based on the experimental data, enabled the insects to skim across the surface to safety faster than those without the adaptation.

5 – Microevolution and Macroevolution

While nature is the selection force for much of evolution, many factors can change the rate at which this occurs. During sexual reproduction, new genes can be added by the random mixing of chromosomes. Mutations from mistakes in meiosis also add genes. Once these new traits originate in a species, the advantageous ones will quickly replicate and spread as nature selects those individuals to survive more, and therefore the frequency of the new genes will increase within the population.

If enough changes occur in the gene pool, reproductive isolation occurs and the result is speciation. The more a population is isolated from other populations, the faster speciation can take place. Also, the smaller the population is, the faster this process happens. Think of it this way: if there are a million finches in a population, then a mutation in one individual has had an impact on only 1/1,000,000 of all of the genes collectively that can be found in the entire population of that species, or the gene pool. It will take many offspring from that one individual to greatly affect the gene pool of that population. However, if the population of finches was only 100, then a mutation in one individual would change 1/100 of the population's gene pool. It would take fewer generations favoring that mutation to cause that trait to become a common feature in that population. As changes accumulate in isolated populations, each population becomes more adapted to its habitat and less like the shared common ancestor. This process of filling different niches and habitats and becoming more diverse through specific adaptations is called adaptive radiation.

Darwin discovered 13 varieties of finches on the Galapagos Islands, but it was not until he returned to England and had the help of an ornithologist that he began to recognize the finches' significance to his incubating ideas about evolution. The ornithologist realized not only that the finches were 13 different species of birds, as Darwin surmised, but also that they were all finches and quite closely related to one another. Darwin eventually concluded that a small population of finches had found their way to the Galapagos Islands, possibly during a storm that blew them away from the mainland of South America. Unable to make the return flight, the marooned population of finches spread throughout the islands. Within the population, there was variation among the beak sizes from small to large. As the population grew and competition for food, space, and other resources increased, those with a beak variation that fit one aspect of the islands survived in those habitats. Birds with beaks better suited for different food sources, such as cracking larger seeds, digging in a cactus, catching insects, or extracting nectar from a flower, became separated by their behavior (seeking different food sources). As populations stopped interacting, they quit reproducing with each other, and the result was 13 isolated populations of finches, which evolved into 13 distinct species of finches, mostly differentiated by their beaks

Researchers Peter and Rosemary Grant have actually observed this evolution of the size of the beaks more than once during 40 years of studying the Galapagos finches. While doing research on the islands, they witnessed a shift in average beak size among the population of medium ground finches as a result to changing weather patterns. First, food became scarce during a severe drought. As seeds ran short, the only seeds left were larger, undesirable seeds that were difficult to crack. The birds with larger beaks had the advantage, and in the following generations the average beak size of the population increased. More individuals with larger beaks survived to produce more, larger-beaked offspring, who themselves then had more, larger-beaked offspring. A few years later, the weather reversed itself and it rained for months on end. This changed the abundance of plant life and therefore the available seeds. Now fortunes shifted, and the birds with the smaller beaks had the advantage in finding the abundant, easy-to-consume, smaller seeds. With this advantage, they reproduced more efficiently, and the next few generations shifted toward smaller-beaked finches within this population (Figure 6.7).

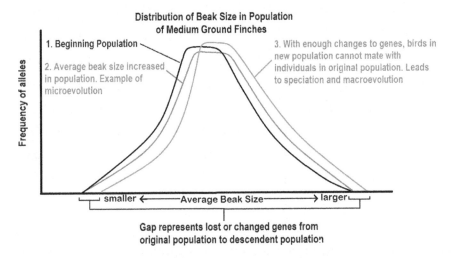

Figure 6.8: Medium Ground Finch Evolution

The Grants were able to collect data more recently during another drought. This time, however, the genetic testing technology existed to map the genome of the finches allowing them to pinpoint the gene HMGA2 as the genetic code responsible for this relatively rapid genetic shift. It is this gene that appears to be the primary genetic coding for beak size. This gene actually controls the collection of other genes that combine for all of the phenotypic characteristics that make up a finches beak.

Such genetic shifting of a single trait in a species is not enough of a change to cause speciation, so it is referred to as microevolution. In this case, the average beak size shifted but all individuals within the population could still mate, so they were still the same species. If the trend continued over enough generations, however-

Video: Micro and
Macro Evolution
*https://vimeo.
com/168782155*

er, the accumulated genetic changes in the birds at the extreme end of the population in regards to beak size could continue to the point where some individuals would be genetically so different than the others they would constitute a new species. If speciation occurs, leading to a new line of organism that can no longer mate with the other population that has evolved from a common ancestor, then this is referred to as macroevolution. When most people think of evolution, it is macroevolution they are visualizing—the genesis of a new species from a previous species. Macroevolution requires an accumulation of many, many genetic changes that would be considered microevolution. This is the reason that evolution takes hundreds of thousands of generations to result in new species that are reproductively isolated from their common ancestor and evolutionary cousins. And because each of us only gets to experience living through one generation, it can be very difficult to comprehend the time involved for macroevolution to occur. But occur it did, as evidenced by countless laboratory experiments simulating genetic changes, natural experiments allowing the observation of genetic changes, and findings in the fossil record.

6 – Our Story

All life in our imagined park and on the Earth, including the human ancestor described at the beginning of this chapter, is the result of adaptive radiation starting with a common, single-celled ancestor as the progenitor to all the diversity of life on the planet. To help make sense of these ancestral relationships, scientists categorize all life on the planet, but the method of categorization constantly changes as scientists discover more about individual species.

The Linnaeus classification system was created with "kingdom" as the largest grouping. Originally, because classification was done by looking at the gross anatomy of the organisms, there were just two kingdoms: plants and animals. As scientists learned more and discovered new species, the system evolved to include five kingdoms: plants, animals, fungi, protists, and monera. Humans are in the animal kingdom. Kingdoms are divided into groupings called phyla. Humans are in the phylum Chordata, which includes all animals with backbones. The phyla are broken into smaller groups called classes. Humans are in the class Mammalia (fur, milk, etc.). Below classes in the taxonomy are the groups of orders, of which humans belong to Primates. Then comes family; humans are in the family Hominidae. The genus (the next level after family) for humans is *Homo*. Species is the last level in the classification system. As you know, humans are *Homo sapiens*. I remember the full classification sequence using the pneumonic device: Kings Play Chess On Funny Green Squares (Figure 6.8).

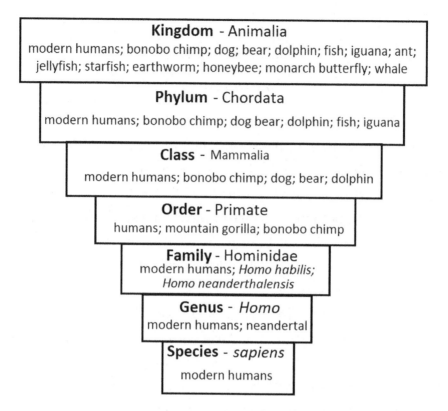

Figure 6.8: Modern Human Classification

As scientists continue to learn more, especially about the genetics and proteins making up the cells, the five-kingdom classification system has been modified; it probably will continue to be modified in the future, as more and more species are examined at the genetic and proteome levels. In fact, most biologists now recognize a new level called domain, which is even higher than kingdom. Domain Bacteria has been added and includes all the modern prokaryotes. Domain Archaea contains the archaic prokaryotes. These are primitive prokaryotes, like those found in deep sea volcanic vents in the ocean. They are believed to be much more similar to the first cells to evolve on Earth than the modern bacteria. The domain Eukarya is divided into four kingdoms that are made up of organisms with eukaryotic cells (the remainder of the original five kingdoms: animals, plants, fungi, and protists).

Central to my own ecological identity is the humbling realization that humans fit on the family tree of life as simply one branch (more like a twig really), off to one side. It is very tempting to think of all other life leading up to us because our "story" is often told in such a way as to end with "and then came humans." The story of evolution is often presented as, "first came the bacteria, then multi-celled creatures, then plants, then fish, then amphibians and reptiles crawling from the water, then animals standing on two feet, and then finally...humans." Or maybe you're more accustomed to the biblical story, where "God created all the creatures out of a void, and finally...man to rule and have dominion." With either version, I believe the danger is the same. Both stories place human beings at the apex of the development of life. Could this method of telling the story be related to the environmental issues with which we are now faced? Is the earth ours to do with as we please? How is this story a part of our ecological identity?

Maybe understanding our evolutionary history will change our perspective on our place in the larger system. I sit here writing these words as a member of the only species left of our genus, *Homo*. Our closest relatives, with which we share approximately 98.5 percent of the same DNA, are the bonobos, and they are not even in the same family. It is unusual for a species to be so isolated. Most species have many more close relatives than we do. Think of the dog genus, *Canis*, for example. It includes many species as a result of natural selection and adaptive radiation: wolves, *Canis lupus*; coyotes, *Canis latrans*; dingoes, *Canis lupus dingo*; and many variations of each of these, as well as all the artificially selected variations (breeds) of the domesticated dog, *Canis lupus familiaris*.

Within the past decade or so, a great deal has been learned by studying the human genome and, in particular, the rate of mutation that occurs within the DNA found in human mitochondria (mtDNA) and the Y-chromosome in males. These two topics are of current interest because both the mtDNA and the Y-chromosome are passed from mother to child or father to son as exact copies. Therefore, as we learn the rate at which mutations occur and catalog what these mutations are within the population of living humans today, we can separate modern humans into different genetic groups (called haplogroups) and get a record of the migration of anatomically modern humans out of Africa. Based on this recent research, called the "Genographic Project," the current thinking is that the modern human species is approximately 150,000 to 200,000 years old based on a variety of fossil and archeological evidence. DNA evidence, specifically the mtDNA carried in all humans and passed from mother to child, can currently be traced back to about 100,000 years ago when we all shared a common maternal ancestor. This "person" that is the maternal ancestor to modern humans is referred to by scientists as Mitochondrial Eve. They are using the image of "Eve" from the Genesis story in the Bible, but they are not saying that this person is *the* Eve from the Bible. The Y-chromosome can be traced to a male about 60,000 years ago and is often referred to as Y-Chromosome Adam (again, not *the* Adam from the Bible).

It is tempting to think of these two individuals as our great-great-great-, etc., grandparents, but notice that they were separated by tens of thousands of years. Instead of imaging them as individuals, we should focus on the genetic material. What it means is that we can trace the mitochondrial genetic material common to all humans back to an individual 100,000 years ago, and the Y-chromosome that is common to all males back to an individual that lived 60,000 years ago. Yes, a single individual did have these chunks of DNA we all share, but we are a mixture of many genes from an entire originating population from Eastern Africa. From Africa, modern humans spread through the Middle East, branched into Asia and Australia, went northward and westward into Europe, and then across modern Russia and into the Americas (Figure 6.9).

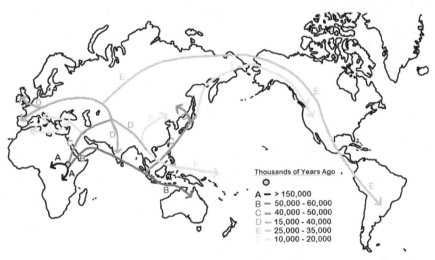

Figure 6.9: Human Migration Based on mtDNA & Y-Chromosome Data from the Genographic Project

One aspect of our story is clear. We are a young species. Yet within the 150,000 – 200,000 years of ours species' existence one readily visible physical variation has evolved, and it is a physical variation that has led to some dark chapters in the human story: skin color. Within that time, the phenotype for skin color has evolved into a broad spectrum of pigmentation. The phenotype of pigmentation is the result of two forms of the pigment melanin: eumelanin and pheomelanin. The different shape of these two molecules means that they are different colors. Eumelanin comes in two types, black and brown. Pheomelanin has a pink or red hue.

This pigmentation has a direct use. Melanin absorbs ultraviolet radiation. The melanin pigment is found in structures called melanosomes which form sort of an umbrella above the nucleus of the cell. This protects the DNA in the nucleus from the UV radiation that can cause structural damage and mutations. The darker the pigmentation, the more UV radiation is blocked. Therefore, it is easy to see that darker skin is an advantage because it can reduce the likelihood of skin cancer. It is hypothesized, based on genetic and fossil evidence that our ancestors lost the majority of body hair and then developed darker skin 1 to 1.5 million years ago. So it would seem that skin pigmentation should be darker in populations that evolved in areas where UV radiation is more intense. The amount of UV radiation is directly proportional to the latitude on the earth. The closer to the equator, the more UV radiation. As you move toward the poles, the UV radiation reduces.

Video: Hominid Migrations https://vimeo.com/168782376

This can be tested by measuring the amount of pigmentation in the skin of individuals from indigenous populations in a variety of locations around the planet. It is necessary to measure the skin pigmentation of individuals with direct ancestry tied to that specific location. In today's world of easy migration, this is not as easy as it sounds. However, Nina Jablonski from Penn

State University was able to collect such data and did find a direct correlation between UV intensity and skin pigmentation. Along with this data, Rich Kittles from the University of Illinois compared the variation in a gene connected to skin pigmentation, the MC1R gene. He found that within individuals indigenous to Africa (with the longest evolutionary history in one location), there was the least amount of variation of this gene. This indicates that there was strong "negative" natural selection pressure on this characteristic—meaning the natural selection pressure was for this trait to remain *unchanged*. But something does not add up here. The obvious advantage to darker skin is protection from skin cancer. However, skin cancer does not usually manifest itself until later in life—after reproductive years for adults. What that means is that while it may be an advantage to a longer life, it is not an advantage during one's reproductive years. We have seen previously that for a trait to be selected for through natural selection it has to be an advantageous trait that leads to producing more and healthier offspring with that trait. Because skin cancer strikes after reproduction, protection from skin cancer is not an advantageous trait, at least in terms of natural selection.

The reproductive advantage connected to skin pigmentation has to do with the vitamin folate. Folate is a needed vitamin in females during pregnancy for healthy embryo development and in males for healthy sperm production. However, strong sunlight breaks down folate. This is the evolutionary advantage for darker skin pigmentation. But if more eumelanin results in melanosomes that protect the cell from skin cancer, and also protects the necessary supply of folate which is a reproductive advantage then why doesn't everyone have dark skin?

The answer to this question provides an example of competing selection factors, which is common in natural selection. Life is often about compromises. Ultraviolet-B radiation is needed by the body to produce vitamin D. Individuals with darker skin actually block most of the UV-B radiation. However, if they live in a climate with very high levels of UV radiation, such as near the equator, they would still get enough exposure, even with the natural protection to the UV radiation, to produce adequate vitamin D. However, populations that evolved in climates further from the equator require more exposure to UV radiation to be able to adequately produce vitamin D, thus the evolution of lighter skin in these populations. This is a compromise, however, because of the lower levels of UV radiation in these latitudes, the body was still able to preserve the needed folate for reproduction. The exception to support this hypothesis is in the population of indigenous peoples living in the Arctic Circle. Despite very low levels of UV radiation exposure, their skin pigmentation is darker than other populations from northern latitudes. The difference is that the Inuit population traditionally survived on a diet that was very rich in vitamin D. Therefore, their bodies got vitamin D from their diet, not by relying on UV radiation to trigger their body naturally producing it and also have retained the advantageous trait of more pigmentation protecting the supply of folate in their cells.

Of course, now the human population is not geographically bound to where their direct ancestors have lived for generations. This means that individuals with very light skin now living in latitudes with high UV radiation need to protect themselves from the sun and conversely individuals with darker pigmentation living in latitudes with low UV radiation may need to adjust their diet to include vitamin D rich foods or take vitamin supplements. One thing is abundantly clear. Skin pigmentation is simply a product of natural selection and nothing else. The production of eumelanin versus pheomelanin is not any indication of other characteristics. Additionally, it is the consensus of geneticists and evolutionary biologists that from a biological standpoint, there is no such thing as "race" in the human population. The variation in genotypes that we see is literally skin deep. Genetically, we are a very young species with very little genetic variation among the population. We are all of one tribe.

But what of the other humans that came before us? A majority of what we know about previous humans comes from the study of fossils and artifacts, and most of these finds are from Africa, Europe, Asia and Australia. Because this is an area of high interest to scientists, new discoveries are being made all the time. With each new discovery, ideas change slightly. Potential new species are added and the story changes. However, there are common, accepted components of the human story among the majority of evolutionary biologists.

Scientists have unearthed more than twenty different species of hominids (primates that walk upright), with more being discovered regularly. Some consist of a single fossil discovery, while others are the results of multiple fossil finds by multiple researchers. Of course, the more fossils found in a variety of locations and by a variety of scientists, the more accepted that species is as a part of the human story.

Our closest relative is *Homo neanderthalensis*. These humans lived from about 230,000 to 30,000 years ago in what is now

Europe. Because much of their existence occurred during times of ice ages, they had a shorter, stockier build than our species as an adaptation to the cold. If you saw *Homo neanderthalensis* in modern clothes, however, you very possibly would not recognize it as a different species. Much like for us today, their culture defined them as much as their biology. They were accomplished hunters; they used stone tools, lived in social groups, and had the beginnings of culture, such as art and ritual, as well as possibly vocal language.

Examination of fossils indicates that they had the same types of traumatic injuries as modern-day rodeo stars, indicating that they hunted large mammals at close contact. This is evidence that these formidable hunters relied on weapons such as spears instead of projectiles. Though they were physically adapted to the cold, they also relied upon behavioral adaptations such as clothing, fire, shelters, and so on to provide insulation from the climate, and in so doing, also insulated themselves from some of the effects of natural selection. That might be why this species, which moved into and then continued living for thousands of generations in a cold climate, had no more fur or hair than modern humans have. A thick protective coat of fur was not a necessary adaptation if you could wear the fur of another animal to survive. Instead, the favored adaptations were those of creativity, innovation, problem-solving and intellect.

This story, then, is as much about the evolution of the brain as it is about the evolution of the body. The early human ancestors had a brain less than half the size of *H. neanderthalensis* and *H. sapiens*. Maybe we should think about the evolution of the brain and cultural adaptations as an example of co-evolution. The brain uses about 25 percent of the energy an animal consumes.

As the hominid brain grew larger, hominid behaviors had to change to feed the hungry brain. This might have triggered a need for higher-calorie food sources such as animal fat and bone marrow. Or maybe the adaptation to develop the hunting skills for taking down larger and larger prey allowed the brain to evolve into a larger, more energy-consuming organ. The need for more energy might have led to the dispersal of the various *Homo* species, and with this dispersal came the cultural adaptations of clothing and building shelter.

We often think of the development of culture as a *revolution,* an event that occurred quickly, shifting humans from living in the "wild" like any other animal to an organism *better* than living in the "wild." Using the paradigm as a revolution can lead one to thinking that when we were "like any other animal" we lived at the mercy of the environment. When food was scarce, we died. When food was abundant, we lived. After the revolution, we became elevated above all of the other animals. This implies a separation from the natural world, and therefore an exemption from the rules that govern all other life on the planet. Maybe instead of a revolution that implies breaking away from nature and its governing ecological and evolutionary rules, we would be well-served to think of our cultural revolution as more of a cultural evolution that has been evolving during the past 30,000 years as a result of natural selection, as a natural part of the biosphere and all of its biodiversity. Within this paradigm, we may still view our species as the most complex, but not necessarily the most "evolved" and certainly not above the rules governing the homeostasis of the biosphere. It is important to make this shift as a matter of survival of our species. As we have seen, an ecological identity that involves enacting a story in which we separate ourselves from nature, and dominating the natural world has not been good for other species, the biosphere, and therefore, finally not conducive to our survival as a species.

This concern over the long term prognosis of our species might explain our fascination with what happened to *Homo neanderthalensis*. Why did our species evolve cultural adaptations to shift from archaic humans to modern humans at about the same time that *Homo neanderthalensis* went extinct? There has been a great deal of debate as to demise of *Homo neanderthalensis*— were they killed by modern humans, did they get absorbed into the modern human population through interbreeding, or were they simply out-competed? One hypothesis is that they simply could not compete with the modern humans that moved into their habitat approximately 50,000 years ago. The competitive exclusion principle states that two species cannot occupy the same niche in a habitat. *Homo neanderthalensis* and *Homo sapiens* most likely were in direct competition for resources in the same niche. Also, the habitat was shifting from ice age to a more temperate climate. These changes and this competition very likely led to, or at least contributed to, the gradual extinction of *Homo neanderthalensis*. Recent genetic research has led some scientists to conclude that their population and genes were absorbed through interbreeding into the *Homo sapien* population. Enough DNA has been collected from preserved bones of *Homo neanderthalensis* to map the genome of this species. Initial findings show that modern *Homo sapiens* do indeed carry some *Homo neanderthalensis* DNA. The exact rate of interbreeding and the connection that interbreeding might have had to the extinction of our closest hominid is yet to be determined. Undoubtedly, the story will continue to change as more is learned.

The longest living member of our genus is a species called *Homo erectus*. This species was most likely the first to migrate out of Africa though some researchers contend this honor goes to another similar species, *Homo ergaster*. *Homo erectus* lived approximately 1.8 million to 300,000 years ago. It used sophisticated stone tools and probably fire as well. *Homo erectus* had a smaller brain than our species and a more ape-like face, but still looked quite human. It is thought that they were well adapted to long-distance running, possibly to run down prey—chasing the prey long enough to exhaust them instead of trying to overpower and kill them like other predators might do. They may have also gotten many of their calories from scavenging.

The oldest known member of the *Homo* genus is *Homo habilis*, (Figure 6.10), which is also the oldest species with which we have found evidence of stone tools. There have been many fossil finds (all located in Africa) that have been classified as *Homo habilis*. Within this species, which dates back from about 1.7 to 2.3 million years, there is a great deal of variation. Some scientists

Figure 6.10 *Homo habilis*

argue that many of these specimens are actually *Homo erectus*, while others belong to the genus *Australopithecus*. This level of variation may suggest that this species is a transition from the early, more ape-like hominids, *Australopithecines*, to the more modern genus, *Homo*.

Since the discovery of Lucy, who we met earlier, many additional samples of *Australopithecines* have been found in Africa, ranging in dates from just over 2 million years old to almost 4 million years old. The fossil record indicates that during this time there was the greatest adaptive radiation of hominids filling a variety of niches in the changing habitat of Eastern Africa. It is likely one of these species is the common ancestor for the *Homo* genus. These are also the oldest specimens with conclusive evidence of upright walking. These creatures were a true mixture of human-like and ape-like characteristics. They were upright walkers, with human-like knees and hips but small brains and very chimp-like skulls and faces.

The *Australopithecines* are probably not the beginning of our story however. Based on the rate of mutation and a comparison of modern human DNA and modern chimpanzee DNA, it is hypothesized that humans and the great apes diverged from each other on the family tree about 6 to 7 million years ago. The oldest fossil found so far, which may be a candidate for our oldest ancestor, is called *Sahelanthropus tchadensis*. This specimen is 6 million years old and has many cranial features that lead scientists to believe it is a possible ancestor. However, insufficient fossil evidence exists at this time to determine if this species was bipedal or not, which is a crucial criterion to be considered a hominid.

We started this chapter with an image of one of our ancestors, an *Australopithecus afarensis*, walking in Africa. From that ancestor, and other species before, our species came about due to the same processes of natural selection that affect the bacteria, butterflies, birds, and all other life on the planet. Going back further yet, we find evidence of a common ancestor uniting all life on the planet. This understanding is the final piece of the puzzle in answering our big idea question regarding our ecological identity.

Web Resource Home Page
www.exploringbiodiversity.com

Chapter six Web Resources
http://www.exploringbiodiversity.
com/#!chapter-6-student-resources/crsp

All ethics so far evolved rest upon a single premise: that the individual is a member of a community of interdependent parts. The land ethic simply enlarges the boundaries of the community to include soils, waters, plants and animals, or collectively the land.

-Aldo Leopold, from The Land Ethic (1949)

Conclusion

Answering the Big Question

The sun is now descending in the sky as we reach the end of our walk. The edge of the park is bordered by a boardwalk that overlooks an ocean-side beach. The ocean horizon is distant, empty, and seemingly endless. The ocean is bigger than you can imagine, yet from space it appears bound and contained on this tiny green and blue planet. Our first stop in developing your ecological identity was an exploration of how individual humans affect the movement of energy and matter throughout an ecosystem. The earth and its biosphere take in energy from the sun, use the energy to create order from the chaos of that energy, and continually recycle the matter into all the abiotic and biotic components of the ecosystem. The individual organism plays a key role in that process, either as a producer, like the plants, or as a consumer, like you. Each species has a specific niche into which it—and only it— fits, and each is connected to the other life forms within its ecosystem. All of these species live together in a community within an ecosystem. Their shared purpose is to maintain the homeostasis of the ecosystem. Changes in the population of one species directly affect species with which it interacts and indirectly affect other species within that ecosystem—and maybe beyond! As an individual within an ecosystem, your body is part of the process of the ecosystem using energy to recycle matter and create order out of chaotic

243

energy. Your body has many intricate mechanisms to maintain its own homeostasis, and in so doing, plays its individual part in maintaining the homeostasis of the ecosystem within which you live.

Your genetic code provides the instructions for how your body maintains homeostasis. It is highly structured and passed on in a very orderly fashion to the next generation. But your life is not determined only by your genes. Even though your genes cannot be changed, your epigenome is constantly interacting with the environment in which you live. This interaction might have an effect on the role genes play in your development, as well as whether a gene is activated or inactivated and to what degree. By looking at our genetic and evolutionary history and understanding how genes interact and are passed from generation to generation, we can begin to understand the history of our species—not just our species, but the other species from which we are descended. Maybe more important, however, is the knowledge that we are genetically linked to all other life on the planet. If we all came from the same ancestry, who's to say we are different or better than any other species? Do the rules of nature that dictate the survival of a species such as an elephant, dinosaur or bacteria also then apply to us?

At one time, we lived in groups determined by blood kin, called clans. Those in the same clan had empathy directed toward understanding and supporting the needs of those closest to them. This was their extended family, and they had a shared interest in each other's survival. As the human population grew and spread, so too did the size of the population to which the individual could extend his or her empathy. Now, beyond just the family clan, individuals belonged to tribes—those with whom they lived and worked to survive by building agricultural societies. The world continued to "shrink" as humans organized into various religious groups, which also expanded the group with which individuals had an affinity.

The advent of easy communication and travel led to affinity groups that consisted of nation-states and even entire percieved races. It is normal to feel empathy towards those in your group; unfortunately it is also normal to feel less empathy for outsiders. These inclinations might lead us to assume that our species is hard-wired for aggression, individualism, and materialism. Certainly, there is ample evidence on the daily news to support this hypothesis. However, our species clearly has empathic sensibilities towards others as seen through the ever-expanding nature of our inclusion groups—from clan, to tribe, to religious group, to nation-states. What if our ecological identity included having empathy not just for members of our own "tribe" but for all those in our species? Or better yet, what if we had empathy for all of the other species on the planet and their right to survival? When viewed this way, and deepening our understanding of our ecological self to include connection to all life, then enacting a story in which we preserve all life is to preserve ourselves. Actions that once felt like sacrifices of lifestyle or comfort, now seem the natural things to do. With this perspective as the foundation of your ecological identity, then, how ought you to live?

Web Resources Home Page *www.exploringbiodiversity.com*

Conclusion Web Resources *http://www.exploringbiodiversity.com/#!blank/cmt9*

Bibliography and Suggested Resources

Aetna Inc. *Tay-Sachs Genetic Testing Basics.* 2011. www.inte-lihealth.com/IH/ihtIH/WSIHW000/32193/35422.html.

American-Isreali Cooperative Enterprise. *Jewish Virtual Library.* 2012. www.jewishvirtuallibrary.org/jsource/biography/calvin.html.

Barlow, Connie. *Green Space, Green Time: The Way of Science.* New York: Copernicus, 1997.

Biomes Group. *The World's Biomes.* Edited by Kacey Ballard. UC Berkeley. 1996. http://www.ucmp.berkeley.edu/glossary/gloss5/biome/.

Brown, J. H., and E. J. Heske. "Control of Desert-Grassland Transition by a Keystone Rodent Guild." *Science* 250 (1990): 1705-07.

Capra, Bernt, Byars, Floyd, and Capra, Fritjof. *Mindwalk.* Directed by Bernt Capra. Produced by Adrianna Cohen. Performed by Ullman, Liv, Waterston, Sam and Heard, John. 1990.

Capra, Fritjof. *The Web of Life: A New Scientific Understanding of Living Systems.* New York: Anchor Books, 1997.

Captrette, David R. *Hans Kreb (1900-1981).* 2005. www.ruf.rice.edu/~bioslabs/studies/mitochondria/krebs.html.

Guns, Germs and Steel. Produced by Lion Television. Performed by Jared Diamond. 2005.

Eichman, Philip. *From the Lipid Bilayer to the Fluid Mosaic: A Brief History of Membrane Models.* n.d. www1.umn.edu/ships/9-2/membrane.htm.

Environment Canada. *Mercury in Food Chain.* 2010.

Envrionmental Protection Agency. *Fish Consumption Advice.*

12 29, 2014. http://www.epa.gov/mercury/advisories.htm.

Frisch, Amos, et al. "Origin and Spread of the 1278insTATC Mutation Causing Tay-Sachs Disease in Ashkenazi Jews: Genetic Drift as a Robust and Parsimonious Hypothesis." *Human Genetics*, 2004: 366-376.

Genetic Science Learning Center. *Learn.Genetics*. University of Utah. n.d.

Goodwin, Timothy. *Exploring Biodiversity*. n.d. www.exploringbiodiversity.com.

Hopwood, Jennifer, et al (2012). *Are Neonicotinoids Killing Bees? A Review of Research into the Effects of Neonicotinoids Insecticides on Bees with Recommendations for Actions*. The Xerxes Society for Invertebrate Conservation.

Howard Hughes Medical Institute: *HHMI Biointeractive* https://www.hhmi.org/biointeractive

Huntington, Henry P. "Using Traditional Ecological Knowledge in Science: Methods and Applications." *Ecological Applications* 10 (2000): 1270-1274.

Institute of Human Origins. *Becoming Human*. 2009. www.becominghuman.org.

Lovelock, James. *The Ages of Gaia: A Biography of Our Living Earth*. New York: Bantam Books, 1988.

Malthus, Thomas. *An Essay on the Principle of Population or A View of Its Past and Present Effects on Human Happiness, 9th Edition*. London: Ballentyne, Hanson and Co., 1888.

Marden, James H, and Melissa G Kramer. "Surface-Skimming Stoneflies: A Possible Intermediate Stage in Insect Flight Evolution." *Science* 266, no. 5184 (October 1994): 427.

Mithen, Steven. *After the Ice: A Global Human History 20,000-5,000 BC*. London: Orion Books, 2006.

National Geographic Society. *The Human Story: Join The Project To Learn About Your Story*. 2015.

National Oceanic and Atmospheric Administration. *PCBs, or Polychlorinated Bophenyls, are Industrial Products or Chemicals*. November 2011.

National Oceanic And Atmospheric Administration. *The Zebra Mussel Invasion*. n.d. http://www.noaa.gov/features/

earthobs_0508/zebra.html.

Neergaard, Lauran. "For the First Time, Scientists Decode Good Bacteria." *Star Tribune (Associated Press)*, 2012.

Nova: Becoming Human. Produced by Graham Townsley. 2009.

PBS Online by WGBH. *Human Numbers Through Time.* 1996-2015. www.pbs.org/wgbh/nova/earth/global-population-growth.html.

PBS/WGBH. *Harvest of Fear.* 2001. http://www.pbs.org/wgbh/harvest/.

—. *Nova.* 2001. http://www.pbs.org/wgbh/nova/body/cracking-the-code-of-life.html.

Peterson, Rolf O. "Wolf-Moose Interaction on Isle Royale: The End of Natural Regulation." *Ecological Applications*, 1999: 10-16.

Pollan, Michael. *The Omnivore's Dilemna: A Natural History of Four Meals.* New York: Penguin Books, 2006.

Ridley, Matt. *Genome: An Autobiography of a Species in 23 Chapters.* New York: HarperCollins Publishers, 1999.

Royal-Woods, Lauren, et al. *The Wacky History of Cell Theory.* n.d. http://ed.ted.com/lessons/the-wacky-history-of-cell-theory.

Shepard, Paul. *Coming Home to the Pleistocene.* Washington D.C.: Island Press, 1998.

Shubin, Neil. *The Universe Within: Discovering the Common History of Rocks, Planets, and People.* New York: Pantheon Books, 2013.

Smil, Vaclav. *Cycles of Life: Civilization and the Biosphere.* New York: Scientific American Library, 1997.

Smithsonian Museum of Natural History. *What Does it Mean to be Human?* April 4, 2015. http://humanorigins.si.edu/research.

Thomashow, Mitchell. *Ecological Identity: Becoming a Reflective Environmentalist.* Cambridge, MA.: MIT Press, 1995.

U.S. Fish and Wildlife Service. *Traditional Ecological Knowledge for Application by Service Scientists.* February 2011.

United States Census Bureau. *Developing Sampling Techniques.* April 23, 2012. www.census.gov/history/www/

innovations/data_collection/developing_sampling_tech-
niques.html.

University of California Museum of Paleontology. *The History of Life: Looking at the Patterns.* n.d. http://evolution.
berkeley.edu/evosite/evo101/IIHistory.shtml.

Vucetich, John A. *The Population Biology of Isle Royale
Wolves and Moose: An Overview.* 2012. www.isleroyale-
wolf.org/.

Weiner, Jonathon. *The Beak of the Finch.* New York: Random
House, 1994.

WGBH/NOVA. *Evolution: A Journey into Where We're From
and Where We're Going.* WGBH Educational Foundation
and Clear Blue Sky Productions Inc. 2001. http://www.pbs.
org/wgbh/evolution/.

Whitt, Sherry, et al. "Genetic Diversity and Selection in the
Maize Starch Pathway." *National Academy of the Sciences,*
2002.

Zhang, X.J., et al. "A Gene for Freckles Maps to Chromosome
4q32-q34." *Journal of Investigative Dermatology,* February 2004: 286-290.

Index

Other Titles from Riverfeet Press:

THIS SIDE OF WILDERNESS (2013)

-Daniel J. Rice

THE UNPEOPLED SEASON (2014)

-Daniel J. Rice

WITHIN THESE WOODS (2015)

-Timothy Goodwin

RELENTLESS (2015)

-Marcus Bruning and Jen Wright

TEACHERS IN THE FOREST (2016)

-Barry Babcock

Bemidji, MN
www.riverfeetpress.com